T0250322

DAMAGE TOLERANCE IN ADVANCED COMPOSITES

DAMAGE TOLERANCE
IN
ADVANCED
COMPOSITES

R. L. Sierakowski, Ph.D.

Professor of Civil Engineering
The Ohio State University

G. M. Newaz, Ph.D.

Professor of Mechanical Engineering
Wayne State University

TECHNOMIC
PUBLISHING CO., INC.

LANCASTER · BASEL

Damage Tolerance in Advanced Composites
a TECHNOMIC⦁publication

Published in the Western Hemisphere by
Technomic Publishing Company, Inc.
851 New Holland Avenue, Box 3535
Lancaster, Pennsylvania 17604 U.S.A.

Distributed in the Rest of the World by
Technomic Publishing AG
Missionsstrasse 44
CH-4055 Basel, Switzerland

Printed in the United States of America
10 9 8 7 6 5 4 3 2 1

Main entry under title:
  Damage Tolerance in Advanced Composites

A Technomic Publishing Company book
Bibliography: p.

Library of Congress Catalog Card No. 94-60957
ISBN No. 1-56676-261-8

To the memories of my wife.
Robert Sierakowski

To the memories of my mother.
Golam Newaz

Preface   *ix*

Chapter 1. Damage Tolerance of Composites 1
  1. Introduction   1
  2. Advanced Structural Composites   1
  3. Damage Tolerance as an Issue   5
  4. Damage Tolerance Definitions   5
  5. Key Elements in Damage Tolerance Concept   6
  6. References   50

Chapter 2. Analytical Methodology 53
  1. Introduction   53
  2. Analysis – Model I   59
  3. Analysis – Model II   64
  4. Analysis – Model III   74
  5. Analysis – Model IV   80
  6. Analysis – Model V   86
  7. Analysis – Model VI   95
  8. Analysis – Model VII   101
  9. Analysis – Model VIII   105
  10. Concluding Remarks   111
  11. References   111

Chapter 3. Damage Tolerance Evalution 113
  1. Introduction   113

2. Nondestructive Evaluation (NDE) Techniques   115
3. Damage Assessment: Optical/SEM Evaluation   131
4. Damage Tolerance Tests   133
5. References   148

Index   151

Advanced composite materials, particularly continuous fiber-reinforced polymers, are currently being used in a wide variety of structural applications. These materials are used not only in the aircraft industry, but in civil, mechanical, and other disciplines in which they are subjected to a wide spectrum of loadings during in-service use. Dynamic loadings, in particular (impact type events), represent a serious design concern for use of advanced aerospace composites for in-service applications as, for example, drop of tools during maintenance. For conventional monolithic materials such as metals, the understanding of damage mechanisms and the assessment of damage occurring from impact events are more widely known and effectively used in design practice. However, when using advanced composites, there is an overall lack of an established knowledge base on how to assess and characterize damage, as well as how to identify the progression of damage and subsequently design against it. All of these issues are related to the structural integrity of the overall system and represent a significant challenge to advanced composite material designers. Accepted design measures for the structural integrity of structural systems using advanced composites include design for strength, stiffness, durability, and safety. The latter measure involves the important issue of this monograph, that is, the issue of damage tolerance. This issue, as exemplified for metallic military aircraft constructions, has been addressed in MIL-A-83444 and for commercial aircraft by manufacturers and the FAA. Damage tolerance requirements for advanced composite structures for general aviation, commercial, and military applications for aircraft are evolving. An important challenge in aircraft design is that the performance of advanced composite material structures will be equal to or superior to the damage tolerance requirements of metallic structures. Thus, an understanding of what damage tolerance is and the role it plays in the use of and design with advanced composite materials for structural applications is the focus of this monograph.

Recognizing that the application of advanced composite materials stems from the use of such materials in the aircraft industry, many of the evolutionary concepts, as related to the damage tolerance issue, pertain to this discipline. Damage tolerance, recognized for centuries in fail safe design concepts, can be traced back to the time of Leonardo da Vinci. However, modern concepts have evolved only within the past several decades.

Chapter 1 discusses the classes of advanced composite material types that are the focus of this monograph, the concepts of damage tolerance that have been discussed primarily in the aerospace community, the important sources of damage that arise in the fabrication and construction of structural components and systems composed of advanced composite materials, and issues that should be considered and addressed by designers. Discussion of damage tolerance of metallic material/structural systems, which include consideration of such defects as flaws and cracks, is presented versus advanced composite material defects that include delamination and disbonding as important issues. The basic philosophy of a damage tolerant definition is presented, including delamination and disbonding as important issues. This includes (1) acceptance that damage will occur, (2) the need for developing an adequate inspection system for damage to be detected, and (3) the requirement that adequate strength be retained within a damaged structure. Finally, the principal drivers related to the design of primary structures composed of composite materials are discussed.

Chapter 2 presents a review of analytical developments associated with the importance of impact damage as a primary driver in the damage tolerance design of composites. Analytical approaches have been catalogued and predictive capabilities and limitations discussed. Chapter 3 presents a discussion of techniques needed for the assessment of damage tolerance and includes a number of important nondestructive evaluation techniques. Illustrative examples of each of the techniques are given along with limitations and applicability. Also, this chapter presents techniques used to evaluate the residual/retained strength of advanced composites as part of the damage tolerance criterion. The importance of retained strength related to the design of the structure and its service life in the presence of nonvisible damage is discussed.

In summary, the purpose of this monograph is to present a current information base to the advanced composite materials community on the subject of damage tolerance, its importance, and its role in the design process. In many respects, it is also a challenge to stimulate further discussion and development of the subject.

Finally, we would like to thank our families, particularly, our wives, Nina until she passed away in February 1995 and Anique for their patience and encouragement during our efforts for this book.

Mr. Manjunath Krishnappa—a graduate student at Wayne State University is acknowledged for his assistance in developing the subject index for the book.

Columbus, Ohio   ROBERT SIERAKOWSKI
Detroit, Michigan   GOLAM NEWAZ
September, 1995

CHAPTER 1

Damage Tolerance of Composites

1. INTRODUCTION

In this opening chapter two issues will be discussed: (1) What is damage tolerance? and (2) Why is this issue a concern? To answer these questions and to develop an understanding of the Damage Tolerance (DT) issue, it is necessary to define the material systems to which a DT definition can be related. Therefore, the first question that is addressed is to define the material systems, which are the subject of this monograph.

2. ADVANCED STRUCTURAL COMPOSITES

Modern structural composites, also referred to as advanced composite materials, are combinations of two or more constituent materials, for example, long stiff fibers embedded in a matrix binder material. The matrix has as its primary function ensuring that the geometrical configuration of the assembly of fibers is maintained, while the long fibers embedded in the matrix generally have a length to diameter ratio in excess of 100. In addition, the strength and stiffness of the fibers may be orders of magnitude greater than the matrix material. Figure 1.1 shows a schematic of the types of composite material systems available, based upon a classification associated with respective constituent reinforcing elements.

Particulate, flake, and random fiber composites are used in specialty applications where stiffness criticality may not be the dominant design driver. Of the types shown, the continuous fiber lamina construction is the most widely used in the aerospace community. Some of the important types of

1

2

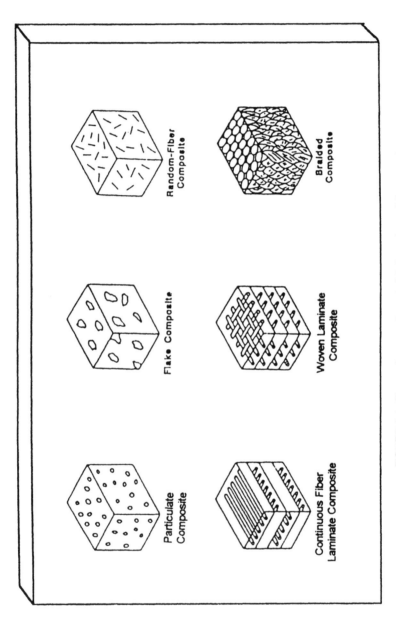

FIGURE 1.1. Types of composite materials based on reinforcement.

MATRIX MATERIALS (Thermosets, Thermoplastics, Metals, Ceramics)		FIBER REINFORCEMENTS	
Acetals	Polypropylenes	Alumina	Al_2O_3
Acrylics	Polymides*	Aluminum	Al
Aluminum*	Polyurethane	Boron*	B
Epoxy*	Phenysilanes	Boron Nitride	BN
Graphite	Titanium	Beryllium	Be
Nickel		Glass*	Gl
Nylon		Graphite*	Gr
Polybenzimidazoles		Aramid (Kevlar)*	Kv
Polyesters		Silicon Carbide	SiC
Polyethylenes		Silicon Nitride	Si_3N_4
Phenolics*		Titanium	Ti
		Tungsten	W
		*principal high-performance reinforcements	

FIGURE 1.2. Advanced composite material matrix and fiber reinforcements.

fiber reinforcements used, as well as matrix systems used, are identified in Figure 1.2.

Fiber reinforcements for use in structural composite construction can be assembled at different levels (Figure 1.3), with a flow chart of the construction types depicted in Figure 1.4.

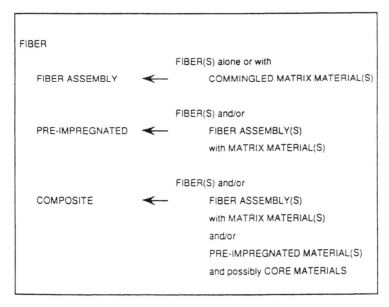

FIGURE 1.3. Levels of composite reinforcement.

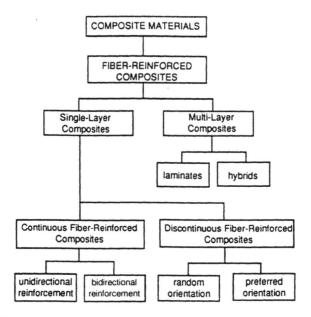

FIGURE 1.4. Flow chart of fiber-reinforced composite material constructions.

The composite material constructions, which are the main focus of this monograph, are shown graphically in Figure 1.5. In addition, advanced composites are generally divided into five categories: (1) polymer matrix composites (PMC), (2) metal matrix composites (MMC), (3) ceramic matrix composites (CMC), (4) carbon/carbon, and (5) hydrids (HC). Even though

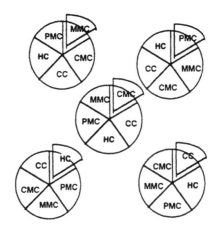

FIGURE 1.5. Types of composite materials.

composite laminates and the types as identified here are the focus of the present discussion, within the configurations shown, there are a number of material systems for which the damage introduced and post-damage performance may require further discussion before quantifying a damage tolerance concept. Thus, a generic concept and definition for damage tolerance of composite structures is difficult to state since such a discussion is dependent upon both the loading and material systems.

3. DAMAGE TOLERANCE AS AN ISSUE

The historical fascination with the subject of flight spawned early interest in what may be described as the first evidence of concern with the subject of damage tolerance (DT) as an important design issue. Indeed, in Leonardo Da Vinci's notebook on the design of flying machines, the following statement is made:

> In constructing wings, one should make one cord to bear the strain and a
> lower one in the same position so that if one breaks under strain, the other
> is in position to serve the same function.

In examining this statement, the built-in redundancy or retaining added structural elements to ensure the structural integrity of the system provides an element of safety to the overall design. With the advent of the demonstrated success of the Wright Brothers in building a flying machine, the evaluation of aircraft construction led to designs that, in the 1960s, incorporated a fail-safe design philosophy for metallic structures [1]. The basic tenet of this approach was that damage introduced by fatigue loading could be detected before the retained strength of the structure was reduced below a safe level. This fail-safe approach was later extended to include the growth of damage associated with both manufacture and/or in-service use. An important point associated with the last statement is that in addition to the concern associated with the fail-safe concept of structural design is the added question of performance of the structure in the presence of and growth of damage.

4. DAMAGE TOLERANCE DEFINITIONS

The concept of damage tolerance can be divided into two stages of evolvement: the pre-1980 approach and that which has emerged since that time. In this regard a number of definitions have appeared in the literature in recent years which include discussions for both metallic and composite materials. A number of these definitions are cited on the next page.

In metallic structures, damage tolerance has been demonstrated using fracture mechanics to characterize crack growth under cyclic loading for the constituent materials, predict the rate of crack growth in the structure under anticipated service loads, and established inspection intervals and nondestructive test procedures to ensure fail safety. [2]

Damage tolerance addresses two distinct types of design philosophies for structures, slow crack growth and the fail-safe issue. [3]

What we really mean by damage tolerance is that we are able to predict the strength of the materials – strength in the broadest sense, i.e., effects of environment, temperature, stress state, damage state, etc. [4]

Generally, damage tolerance is concerned with the ability of the structure to contain representative weakening defects under representative loading and environment without suffering excessive reduction in residual strength, for some stipulated period of service. [5]

Damage tolerance assumes the existence of initial flaws in the structure and the structure is designed to retain adequate residual strength until damage is detected and corrective actions taken. [6]

The structure must be designed in such a way that any damage incurred from normal operation is detectable before the strength or stiffness of the structure falls to an unacceptable level. [7]

The ability of the airframe to resist failure due to the presence of flaws, cracks, or other damage for a specified period of time. [8]

Damage tolerance for aerospace composites is defined as the capability of the composite structure to sustain an impact event with barely visible damage and retain appropriate residual strength. [9]

The current damage tolerance philosophy states that a structure should sustain lifetime design loads in a damaged condition up to that at which barely visible damage is detectable. [10]

The amount of damage just before causing a failure of the structural component represents the degree of damage tolerance. [11]

5. KEY ELEMENTS IN DAMAGE TOLERANCE CONCEPT

As observed from the preceeding definitions, the subject of developing damage tolerance requirements for material/structural systems is an evolving process. For example, in the case of aircraft design, damage tolerance requirements are stated in MIL-A-83444, while for commercial aircraft, the manufacturer and the Federal Aviation Agency (FAA) state the requirements. The damage tolerance requirements for composites are yet to be established; however, a basic tenet in the design of structures composed of composite materials is that the structure should equal or exceed the damage tolerant requirements of metals. This requirement, along with the damage tolerant definitions cited in Section 4, raises a number of issues related to the formulation of a damage tolerance definition for composites. Among these issues are

(1) Quantifying measures of damage for metals versus composites

(2) Specifying the level of design complexity, for example, material versus structural

(3) Identifying the relation of damage tolerance as a design issue and its relation to durability and structural integrity

5.1 Measures of Damage: Metals versus Composites

In the case of metals versus composites, a key overriding issue related to damage tolerance is the difference in the formation of damage and the resulting damage mechanisms. For metals, damage can be introduced by fabrication flaws, scratches, and by in-service use including such effects as cracks and dents. In the case of composites, specifically laminated composites, delamination and disbonds are commonly observed as visible damage mechanisms. While the classical methods of fracture mechanics may be applicable to metals for developing measures of damage, the fundamental differences in characterizing damage mechanisms in composites requires new methodologies to be developed. Thus, it may be stated that methodologies for evaluating damage measures in metals are fairly well established, and are in the process of being established for composites. For example, the damage tolerance of metallic structures is specified in MIL-A-83444 [12].

The damage tolerance design requirements, as specified in MIL-A-83444, apply to the safety of all flight structures, i.e., structures whose failure could cause direct loss of the aircraft or whose failure (if it remained undetected) could result in the loss of aircraft. The requirements stipulate that damage is assumed to exist in each element of a new structure in a conservative fashion (i.e., a critical geometric orientation of the structure with respect to the applied stress field and at a region of highest stress). The structure is required to contain the growth of the initial assumed damage for a specified period of service while maintaining a minimum level of residual static strength, both during and at the end of this period. Figure 1.6 illustrates these requirements in a schematic form. Since residual static strength generally decreases with increased damage size, the residual strength and growth requirements are coupled through the maximum allowable damage size, i.e., the damage size growth limit established by the minimum required residual strength load. The safe growth period (period of unrepaired

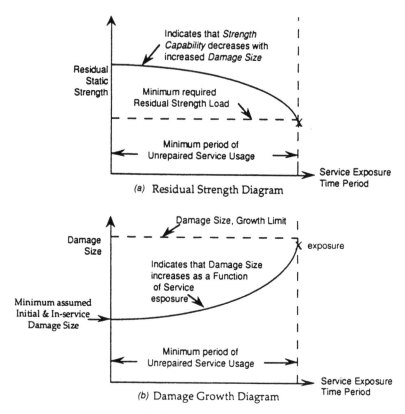

FIGURE 1.6. Residual and damage growth diagrams.

service usage) is coupled to either the design life requirement for the air ve-
hicle or to the scheduled in-service inspection intervals. While the specific
requirements of MIL-A-83444 may seem more complex than described in
Figure 1.6, all essential elements are illustrated.

A structure can qualify for damage tolerance design acceptance for the
case of metallic structures and according to MIL-A-83444 under one of two
categories:

- *slow crack growth* – structures that are designed such that initial
  damage will grow at a stable, slow rate in a service environment
  and not achieve a size large enough to cause rapid unstable crack
  propagation
- *fail safe* – structures that are designed such that propagation damage
  is safely contained by failing a major load path or by other damage
  assessment features

In the slow crack growth category an acceptable damage tolerance for the structure (and thus safety) is assured only by the maintenance of a slow rate of growth of damage, a residual strength capacity, and the assurance that subcritical damage will either be detected at the depot or will not reach unstable dimensions within several design lifetimes. In the fail safe category, an acceptable damage tolerance for the structure (and thus safety) is assured by the allowance for partial structural failure, the ability to detect this failure prior to total loss of the structure, the ability to operate safely with the partial failure prior to inspection, and the maintenance of a specified static residual strength throughout this period.

The damage-tolerant design requirement addresses both the residual strength issue and the damage propagation issue for the structure under consideration.

The damage growth issue can be displayed graphically by plotting damage size as a function of time. The period of time required for damage to grow from a minimum detectable size, such as by in-service inspection, to a critical size is the detection period. The critical size generally corresponds to the maximum service load residual strength, although for composites, this value could be dictated by a stiffness loss to a critical value.

The residual strength is the amount of static strength available at any time during the service exposure with damage present. Safety is ensured by designing requirements wherein damage is never allowed to grow and reduce the residual static strength of the structure below a specified value, such as that corresponding to the maximum load to be experienced in service.

These factors can be construed to imply a damage tolerance definition, that is, damage tolerance is concerned with the ability of the structure to contain representative weakening defects under representative mechanical, thermal, and environmental loading without suffering excessive reduction in residual strength for a stipulated period of time. A graph of the damage tolerance issue is presented in Figure 1.7.

A careful review of these issues indicates several factors that are inherently important in the development of a damage tolerant design criterion:

- the acceptance that damage is present or will occur
- an adequate system of inspection so that damage may be detected
- an adequate strength retained in the damaged structure

These factors are graphically depicted in Figure 1.8.

The key elements and the major variables typically investigated to develop a damage tolerance methodology are described chronologically in Table 1.1. The corresponding evaluation approaches are also described.

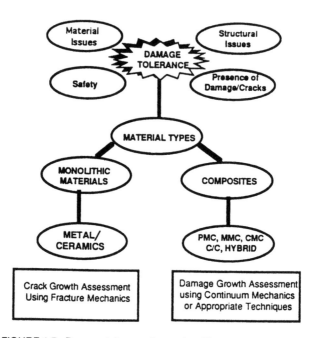

FIGURE 1.7. Damage tolerance issues for different material systems.

5.2 Role of Failure Mode in Damage Tolerance Criteria

Two critical elements of the damage tolerance concept are (1) identification of the most severe and common form of loading event, which results in damage that is structurally unacceptable, and (2) identification of the resulting failure mode associated with that damage. For polymeric composites, low velocity transverse impact has been identified to be one of the most common forms of loading event (typically occurs during servicing the aircraft) that results in delamination (failure mode), which can adversely affect the structural integrity of the composite structure. The structural capacity is assessed typically by measuring the residual properties that are of significant design interest for the specific application of interest. For laminated polymeric composites used in aircraft structures, compression strength is considered to be a critical measure as it is strongly influenced by ply delaminations and is quite relevant in structural design. To put it simply, the damage event and its consequence on the retained structural capacity of the composite should not fall below a certain level such that the composite functional capability is impaired. Therefore, the concept of damage tolerance and its implementation is a way to ensure that the composite structure can be used without compromising safety during normal service.

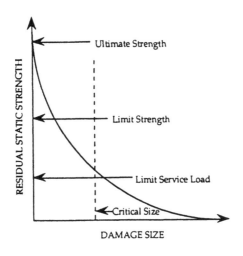

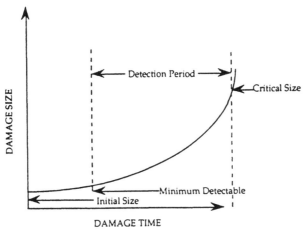

FIGURE 1.8. Damage growth and residual strength.

TABLE 1.1. Key Elements for Damage Tolerance Methodology.

Key Element	Major Variables	Evaluation Approval
Initial loading	Flexural moment (flexural energy) impact energy, critical shear, and normal stresses	Quasi-static flexure, flexural impact
Damage inspection and characterization	Damage size, location, and distribution	Video-microscopic on-line, evaluation of C-scan NDT
Failure mechanism	Damage initiation and growth. Damage type and source (normal stresses, shear stresses, etc.)	Acoustic emission fractograph
Stress analysis	Stress/strain distribution at damage vicinity. Equivalent stress for critical level (propagation or failure)	Interlaminar 2D analysis Interlaminar 3D analysis
Post-damage loading	Cyclic loading vs. damage growth Number of cycles vs. damage growth	Quasi-static compression, compression-tension, compression-compression cyclic loading.
Residual characterization	Residual compressive strength, stiffness Residual fatigue life	Quasi-static compression cyclic loading
Design and optimization	The effect of layer sequence The effect of tough interlayers The effect of hybridization	Tension, compression, flexure 2D loading

The damage tolerance basis for PMCs is fairly strong in that knowledge and experience has allowed for the development of the appropriate concepts to mature and to be tested. These concepts can be collectively defined as "Compression-After-Impact" (CAI) criteria. The damage tolerance concepts for other types of composite material and structural systems such as metal matrix composites (MMC), ceramic matrix composites (CMC), Carbon/Carbon, or hybrid systems are not well defined or mature. For these new material or structural systems, CAI may not be appropriate and, therefore, key elements to develop a rational plan for damage tolerance for these new systems must be defined.

As an example, let us explore the critical issues for MMCs for damage tolerance in order to examine the requirements for a new material system. These composites are primarily designed for thermostructural applications, although aerostructural applications are also anticipated for limited cases. In an engine application, impact is not the main concern; rather thermal and thermomechanical loads are common. Delamination is not a prevalent failure mode in laminated MMCs. Furthermore, MMCs are most likely to be used in the unidirectional fiber orientation in structures that favor this configuration such as in rotors, blades, and disks. For unidirectional MMCs, a major weakness is the fiber/matrix debonding due to applied service loads, which is further aggravated due to the difference in thermal expansion coefficient between the fiber and the matrix. Let us now draw an analogy for PMCs with the key elements for damage tolerance for MMCs. For MMCs, severe loading: thermal and thermomechanical loading (as opposed to impact loading), leads to a failure mode: fiber/matrix debonding or fiber fracture as opposed to delamination). The effect of debonding is likely to degrade a number of other properties. However, from a structural design standpoint, the tensile residual strength (as opposed to compression strength) would be a logical design choice since this property is more relevant to MMCs in most applications.

For PMCs, the damage tolerance concepts are mature in that failure modes for specific applications are clearly identified. Damage tolerance guidance for aircraft having PMC structural components is defined by certain characteristic flaw assumptions due to manufacturing and in-service conditions. These are listed in Table 1.2. Delamination is identified as a major failure mode in these considerations.

Damage tolerance can thus be addressed at several levels of design complexity. Considering damage tolerance at the material level, such as PMC composites, tough resins and high strain to failure fibers are desirable features. Indeed, added improvement can be achieved through the containment of damage within an acceptable region and through the use of microstructure to arrest damage growth. At the global/macrostructural level,

TABLE 1.2. Manufacturing Initial Flaw/Damage Assumptions.

Damage/Flaw Type	Damage/Flaw Size
Scratches	Assume the presence of a surface scratch that is 4.0" (10.2 cm) long and 0.02" (5.1 cm) deep.
Delamination	Assume the presence of an interply delamination that has an area equivalent to a 1.0" (2.54 cm) diameter circle with dimensions most critical to its location.
Impact Damage	Assume the presence of damage caused by the impact of a 1.0" (2.54 cm) diameter hemispherical impactor with 100 ft/lb (135.6 J) of kinetic energy or with that kinetic energy required to cause a dent 0.10" (0.254 cm) deep, whichever is least.

damage tolerance can be addressed through load redistribution following a regional/local failure. Thus, a number of improvements can be achieved through advances in improved matrix materials, damage containment, and design load redistribution.

5.3 Level of Design Complexity: Material versus Structural

The use of carbon fiber composites in the design of secondary structures in aircraft has become well established during the last decade. Such materials are also now being applied to the volume production of primary structures, especially on military aircraft. The interest to develop this technology for large primary structure in civil aircraft is very high as a result of the incentive for weight savings.

TABLE 1.3. In-Service Nonflight Damage Assumptions.[1]

Damage/Flaw Type	Damage/Flaw Size[2]
Scratches	Assume the presence of a surface scratch 4.0" (10.2 cm) long and 0.02" (5.1 cm) deep
Impact Damage	Assume the presence of damage caused by the impact of a 1.0" (2.54 cm) diameter hemispherical impactor with 100 ft/lb (135.6 J) of kinetic energy or with that kinetic energy required to cause a dent 0.10" (0.254 cm) deep, whichever is least.

[1]The damage listed is assumed to be noninspectable.
[2]Identifies areas of the aircraft where there is a reduced possibility of threat and ensures that the assumed damage area is reduced.

Epoxy matrix composites have several very attractive features. These resin systems are compatible with graphite fibers, thereby eliminating many of the interface problems that may be evident in the newer type thermoset and thermoplastic systems. They are also resistant to aircraft fluids such as jet fuel or hydraulic fluid. Finally, large data bases and long flight histories exist for these systems both in military and commercial aircraft. The first generation epoxy systems are, however, brittle and limited to relatively low design strain levels. This, coupled with the vulnerability of these materials to foreign object impact damage, defines the need for improved toughened composites. If new composite systems could be developed that exhibit greatly improved toughness characteristics, higher allowable design strains of 0.005 to 0.006 would be feasible, thereby offering the potential for even greater weight savings over aluminum and greatly improving the commercial potential for such new materials.

To be effective, the next generation of composite material systems must be designed to provide the requisite in-plane structural properties (stiffness and strength) and to increase durability and damage tolerance. The material's use will be severely limited if a balance in these properties is not maintained.

5.3.1 DESIGN-RELATED MATERIAL AND STRUCTURAL PERFORMANCE

The discussion and figures in this sub-section is adapted from Evans and Master's article in ASTM publication [13], "A New Generation of Epoxy Composites for Primary Structural Applications: Materials and Mechanics," reprinted with permission from *STP 937—Toughened Composites*, Copyright 1987 American Society for Testing and Materials, 1916 Race St., Philadelphia, PA 19103.

5.3.1.1 Hot/Wet Compressive Strength

The material system must, above all else, meet the structural requirements of the aircraft while economically providing a meaningful weight savings over aluminum. A quantitative factor determining the amount of weight saved can be calculated by comparing the specific strengths of the quasi-isotropic composite laminates with aluminum. The quasi-isotropic laminate configuration is chosen because it is representative of in service structural component configurations. A valid comparison must also consider environmental conditioning and loading type. The matrix dominated hot/wet compressive strength, considered the severest measure of structural performance, is used for this comparison. The test temperature is

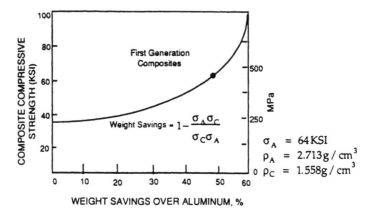

FIGURE 1.9. Quasi-isotropic laminate hot/wet compressive strength used as a measure of structural performance.

defined by material application: civilian and transport aircraft require a 93°C upper temperature range; some military aircraft applications require performance at 132°C.

Based on these considerations, weight saving is defined by Equation (1). A graphic representation of this equation is presented in Figure 1.9. Assuming a design ultimate strength of 440 MPa (64 ksi) for 2024 T3 aluminum and a 60% fiber volume in the composite material, a quasi-isotropic compressive strength of 360 MPa (52 ksi) under worst case hot/wet conditions would be needed to obtain a 30% weight saving over aluminum. As the figure indicates, typical first generation graphite/epoxy systems (tested at 90°C/wet) easily meet this weight saving criterion.

$$\% \text{ Weight Saving} = \frac{\varrho_A - \left(\frac{\sigma_A}{\sigma_C} \cdot \varrho_C\right)}{\varrho_A} \times 100 \qquad (1)$$

where

σ_A = compressive strength allowable Al = 64 ksi (440 MPa)
ϱ_A = density Al = 2.713 gm/cm³
σ_C = compressive strength composite
ϱ_C = density composite

Note:

$$\frac{1}{\varrho_C} = \frac{K}{\varrho_R} + \frac{(1 - K)}{\varrho_f}$$

ϱ_f = fiber density = 1.76 gm/cm³
ϱ_R = resin density = 1.26 gm/cm³
K = resin weight fraction

In order to evaluate structural performance sixteen-ply-thick laminates with a $[\pm 45/0/90/0/\pm 45]_s$ stacking sequence were used to monitor structural performance in compression. The compressive strength specimens were 80 mm (3.15 in.) long and 12.7 mm (0.5 in.) wide; 38 mm (1.5 in.) with long tabs bonded to the specimen, effectively reducing the test section length to 4.78 mm (0.188 in.). The specimens were loaded in a 90 kN (20,000 lb$_f$) Instron test machine at a deflection rate of 1.27 mm/min (0.05 in./min). Side supports similar to the support jig used in the ASTM Test for Compressive Properties of Rigid Plastics (D695) were used to prevent gross Euler buckling during loading. The support fixtures contact the tabs only, leaving the test section unsupported.

5.3.1.2 Impact Damage Resistance

Although it is more difficult quantitatively to assess the impact resistance of a composite material, it is possible to measure relative performance. A variety of test techniques have been employed to measure foreign object impact behavior. The extent of damage has been reported in terms of damage area after a drop weight impact [14], through penetration impact using an instrumented impact fixture [15], impact while under a compressive load [16], and residual compression after impact [17]. All tests, however, produce only comparative values of one system against another. Just as in the actual aircraft structure, the damage resulting from a given blow will depend on a number of factors such as local rigidity, energy absorbed by the support structure, shear stress versus flexural stress, and so forth.

The data in this presentation are based on the impact test procedure described by Byers [18]. These tests use specimens that are 152.5 mm (6 in.) long, 100 mm (4 in.) wide, and 36 plies thick. They have a $[(\pm 45/0/90/0/90)_2/\pm 45/0/90/\pm 45]_s$ stacking sequence. The specimens were positioned on a steel plate having a 76- by 127-mm (3- by 5-in.) rectangular cutout, which supported the specimens along their edges. A 1.8-kg (4-lb) weight was dropped down a calibrated tube, striking a 1.6-cm (0.62-in.) diameter spherical head impactor that rested on the specimen. While various impact energies can be applied, 680 kg m/m (that is, 1500 in.-lb/in. of laminate thickness) was chosen as an arbitrary standard. The support, size, and thickness of the specimen gave an effect similar to dropping a tool from a few feet onto an aircraft wing and striking it between stiffeners.

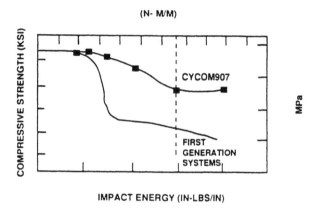

FIGURE 1.10. The reduction in laminate residual compressive strength increases at higher impact energy levels.

Subsequent to impact, test panels were loaded to failure in a 225-kg (50,000-lb$_f$) Instron test machine. A test fixture that provides simple support along the specimen edges is used to prevent gross buckling during loading.

The usefulness of this test in assessing toughness can be seen when one compares the compressive strength after impact data for CYCOM®907, a modified epoxy that is often used as a benchmark for toughness, and a typical first generation epoxy system. Figure 1.10 plots these residual compressive strengths over a range of impact energies. At low impact energies, there is little difference between the two systems since both are capable of withstanding these impact energies without sustaining damage; that is, both have good impact resistance. However, as the impact energy increases, the situation changes dramatically with the tough system being much more capable of withstanding higher impact without damage. When the tough system is damaged, there is a smaller effect on the residual compressive strength; that is, it is more damage tolerant. From the data presented in Figure 1.10, it can be seen that the compressive strength after impact (CAI) of CYCOM®907, namely 280 MPa, is double that achievable with first generation materials at the 680–kg-m/m (1500–in.-lb/in.) impact energy level.

5.3.1.3 Matrix Requirements for Toughened Composites

The problem of concern is illustrated in Figure 1.11, where hot/wet performance is plotted versus compressive strength after impact. As can be seen, the first generation materials have good hot/wet performance, but very low compressive strengths after impact. Alternatively, while

CYCOM®907 has good compressive strength after impact, it is far below the 30% weight savings target. This is typical of the types of trade-off made between hot/wet performance and improved damage resistance and tolerance. It is necessary, therefore, to address the development of a new generation of epoxies having the balance of properties required. This can be accomplished by first examining the material failure mechanisms for compressive loading and impact loading.

In-plane compressive loading is a severe and limiting test of the material. Several potential failure modes exist at practical fiber volume fractions, that is, for $V_f > 0.40$ [19]. They include (1) transverse tensile failure, (2) fiber microbuckling, and (3) shear failure. Although failure is controlled by the matrix resin, the situation is greatly complicated by the tendency of the fibers to bend and buckle. Axial compression produces shear load components between the fiber and the matrix because the fibers are not perfectly aligned within the ply. These out-of-plane components can induce tensile loads in the matrix, which can lead to matrix yielding, matrix microcracking, or fiber-matrix debonding. These latter effects precede transverse failure and fiber microbuckling and may cause premature structural failure.

This model is consistent with a British Aerospace study [26] that indicated that, in general, for an epoxy matrix reinforced with a highly anisotropic fiber, failure is likely to occur by microbuckling [21]. Other studies [22,23] have shown that, in composites with soft matrices (matrix shear moduli less than 690 MPa), compressive strength is strongly influenced by resin modulus. Analogous results were reported for boron/epoxy composites [24]. While more precise empirical relationships can be developed, the

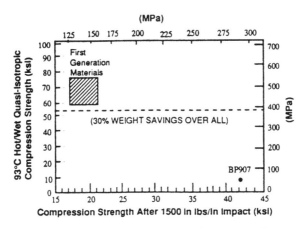

FIGURE 1.11. A trade-off is made between hot/wet performance and improved impact resistance.

neat resin shear modulus and flexural modulus were measured under hot/wet conditions can be used as a first approximation to determine the composite compressive strength [see Equation (2)].

$$\sigma_{cc} = K_1 G_R = K_1 \left[\frac{E_r}{2(1 + \nu)}\right] \qquad (2)$$

where

σ_{cc} = composite compressive strength
K_1 = constant
G_R = resin shear modulus ≤ 689.4 MPa (100,000 psi)
E_r = Young's modulus of resin, psi
ν = Poisson's ratio of resin

In addition, the importance of the fiber matrix interface bond is demonstrated by the direct relationship between compressive strength and interlaminar shear strength that has been seen for some materials [25,26]. Its effect on failure is also important [27,28], and any degradation at the fiber matrix interface causes a falloff in compressive strength [29].

A quantitative assessment of the factors important in the initiation of damage from foreign objects has been presented by Dorey [30]. He conducted studies that indicated that, when carbon fiber–reinforced plastics are subjected to transverse impact, the type of damage that occurs depends on the incident energy and momentum, material properties, and the geometry. No damage occurs if the energy of the striker is accommodated by the elastic strain energy in the material. He calculated the energies to cause:

delamination: $(2/9)(\tau^2/D)(wl^3/t)$ $\qquad$ (3)

flexural fracture: $(1/18)(\sigma^2/E)(w/t)$ $\qquad$ (4)

penetration: $\pi\gamma td$ $\qquad$ (5)

where τ is the interlaminar shear strength; σ the flexural strength of the composite; E the Young's modulus of the composite; γ the through-the-thickness fracture energy; d the diameter of the projectile; and w, l, and t the width, length, and thickness, respectively, of the flexed part of the test specimen. Whether delamination or flexural fracture occurs, depends on the relative values of τ and σ and the span-to-depth ratio l/t; impact damage is less likely when there are low modulus layers on the outside of the specimen such as $\pm 45°$ layers of Kevlar or glass fibers. Whether penetration occurs, depends not only on the incident energy, but on the size and shape of the striker; penetration is more likely for small masses traveling at high velocities (see Figure 1.12).

Dorey's analysis is based on linear elastic modeling and does not examine what happens when the critical values for initiation of damage are attained. If a crack is started, this crack will grow until the stored energy is dissipated. In brittle systems, this will result in crack growth as shown in Figure 1.4. However, if the system has the ability to yield and undergo plastic deformation, this energy will be dissipated without damage growth. Since graphite fibers exhibit a linear elastic failure, improvements must be made in the matrix resins to increase significantly laminate impact resistance and damage tolerance. The matrix characteristic that is most critical to improving composite toughness is the ability to sustain a high stress while yielding, that is, to develop a "knee" in the stress-strain curve and have a large strain to failure. This is clearly evident when the shear stress-strain curve of a first generation matrix and CYCOM907 are compared, as shown in Figure 1.13.

While Dorey's equations indicate that lowering the bending modulus is another means to increase resistance to delamination resistance, a large in-plane modulus is required to control fiber microbuckling and damage growth under compressive loads. Reductions of the composite bending modulus should only be made by changing the laminate lay-up sequence and not through reductions in resin modulus. In fact, it is beneficial to overall performance to increase the resin modulus if the shear strength and flexural strength are also increased and to keep the ratios τ^2/E and σ^2/E high.

As a first approximation for single-phase epoxy systems, the area under the neat resin's stress-strain curve, that is, the work to failure, is assumed to be proportional to the compressive strength after impact. In particular, the

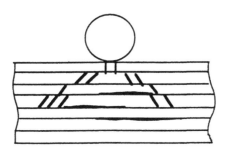

FIGURE 1.12. Low velocity impact causes internal delamination in brittle resin systems.

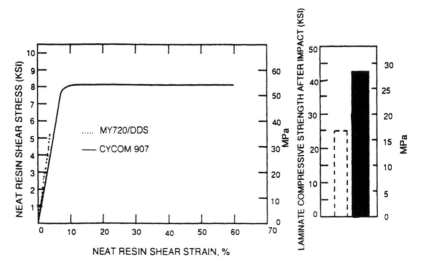

FIGURE 1.13. Neat resin shear stress–strain response is indicative of composite toughness.

existence of plastic deformation will be the major contributor to the work to failure. This may be expressed mathematically by:

$$CAI = K \int_0^{\epsilon_f} \sigma(\epsilon)d\epsilon$$

where

CAI = compressive strength after impact
K = constant
σ = stress
ϵ = strain

The area under the stress-strain curve can be measured using either a flexural test, such as described in ASTM Tests for Flexural Properties of Unreinforced and Reinforced Plastics and Electric Insulating Materials (D 790), or in shear using the Iosipescu fixture (ASTM D 790 was used in this evaluation). For single-phase systems, the same type of stress-strain behavior will be seen in both tests. For multiphase systems, however, a material can undergo high plastic shear deformations without showing similar behavior in a flexural or tensile mode.

A major effort in the last fifteen years to improve damage tolerance has been focused on developing new toughened matrices to significantly improve CAI. This area requires special consideration and is elaborated in the next section. Why toughened systems are desirable and how they improve overall damage tolerance is a subject of interest by itself.

5.3.2 NEW COMPOSITE MATRICES

Most first generation epoxy matrix resins are based on the MY720 (tetraglycidal diamino diphenyl methane) epoxy resin cured with diamino diphenyl sulfone (DDS), either with or without a boron trifloride·monoethylamine (BF3·MEA) catalyst system. Such matrix resins have good hot/wet performance but are inherently brittle, as depicted by their resin properties shown in Figure 1.14. Attempts to modify these systems through the addition of rubber-type compounds or elastomers, or both, show only a nominal improvement in toughness. More importantly, the hot/wet performance of systems modified in this matter is drastically reduced, making them no longer applicable for use in primary structures. Similarly, diglycidal ether of Bisphenol A (DGEBA) resins cured with dicyandiamide (DICY), as shown in Figure 1.15, have very good strain to failure and exhibit the desired type of plastic deformation, but, again, the hot/wet performance at 90°C is almost zero.

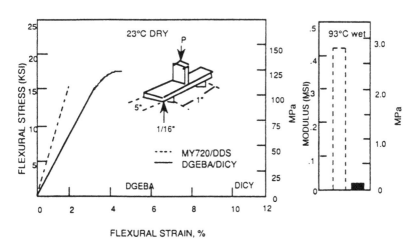

FIGURE 1.14. Flexural stress–strain properties for brittle and rubber toughened neat matrix resins.

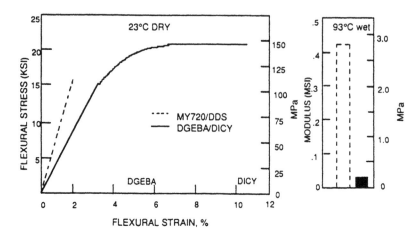

FIGURE 1.15. Comparison of flexural stress–strain behavior of brittle and tough neat resins.

In order to achieve a significant improvement in resin strain to failure, while at the same time retaining a high wet glass transition temperature (T_g) and modulus, a strategy was developed based upon continuing to use MY720 as the base resin but in combination with newly synthesized flexible diamine curing agents, novel catalyst systems, and polymeric modifiers. This new curative chemistry allowed strain-to-failure values of the matrix resins to be improved from the typically 2.5 to 3.0% level up to values greater than 10%, as shown in Figure 1.16. These new systems have shown

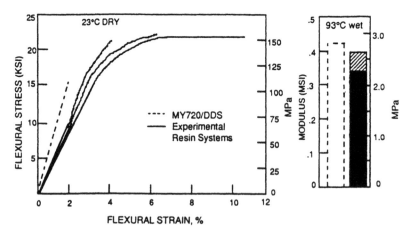

FIGURE 1.16. Comparison of flexural stress–strain behavior of brittle and experimental toughened neat resins.

the required toughness concept of a "knee" in the stress-strain curve – that is, the ability to sustain a high load while "yielding."

The transition to composites performance with these new systems is very good. As shown in Figure 1.17, a linear relationship exists between the neat resin flexural strain to failure and the composite's transverse tensile strain to failure [measured using the ASTM Test for Tensile Properties of Fiber-Resin Composites (D 3039)]. Furthermore, these resins have demonstrated a very good ability to wet the fibers.

More importantly, as shown in Figure 1.18, this transition of enhanced resin properties to composite laminate performance holds for compressive strength after impact. The next generation epoxy systems with strains to failure in the 8% range, for example, CYCOM®1806 and CYCOM®1808, have compressive strengths after impact comparable to CYCOM®907. First generation materials with a strain to failure of 2.5%, on the other hand, have only a 138-MPa (20 ksi) residual compressive strength.

Hirschbuehler [31] presented a more extensive comparison of Cyanamid's data on the relationships between resin chemistry and composite performances. In turn he noted a similar correlation between neat resin ultimate strain and compressive strength after impact.

Finally, the retention of the hot/wet resin modulus (as shown in Figure 1.19) results in a laminate hot/wet compressive strength required to meet the established weight savings criteria. Plotting data in a manner similar to that shown earlier, it can be seen that the new generation of single-phase epoxy systems has remained above the 30% weight savings line and moved far out to the right on the plot, indicating their suitability for use in primary structure applications. It is also shown that the neat resin data included in (Table 1.4) is a good indicator of composite laminate performance (Table 1.5).

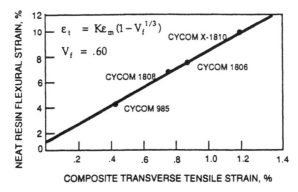

FIGURE 1.17. Correlation of neat matrix resin flexural failure strain and composite transverse failure strain.

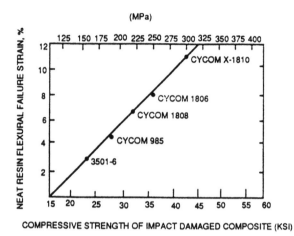

FIGURE 1.18. Correlation of flexural strain-to-failure to neat matrix resins and residual compressive strength of impacted laminates using those resins.

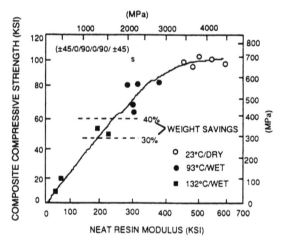

FIGURE 1.19. Relationship between neat matrix resin flexural modulus and quasi-isotropic laminate compressive strength.

TABLE 1.4. Comparison of Neat Resin Properties.

Resin Property	MY-720 DDS	CYCOM 907	CYCOM 985	CYCOM 1806	CYCOM 1808	CYCOM X-1810
				Material		
Flexural modulus, Msi	0.601	0.470	0.590	0.530	0.487	0.494
Flexural strength, ksi	15.2	18.8	22.5	23.7	25.5	24.0
Flexural strain, %	2.5	5.0	4.5	8.0	7.3	>10
Work-to-break, in.-lb/in.	205	—	570	1375	1250	>1800
σ^2/E	0.384	0.752	0.858	1.060	1.335	1.166
Flexural modulus, Msi, at 90°C/wet	0.400	0.01	0.380	0.330	0.325	0.343
Morphology	single phase	multi phse	single phase	single phase	single phase	single phase

TABLE 1.5. Comparison of Composite Properties.

Resin Property	Material[1]					
	MY-720 DDS	CYCOM 907	CYCOM 985	CYCOM 1806	CYCOM 1808	CYCOM X-1810
Quasi-isotropic						
Compressive strength						
23°C dry	100	91	100	104	97	105
93°C dry	100	55	95	90	91	85
93°C wet	88	9	71	75	86	65
132°C dry	88	–	63	73	87	65
132°C wet	58	–	51	–	56	–
Transverse tensile modulus, Msi	–	–	1.20	1.17	1.12	1.15
Transverse tensile strength, ksi	–	–	6.0	9.7	8.1	12.0
Transverse tensile strain, %	4.2	–	0.50	0.86	0.72	1.16
Damage area after impact, in.[1]		0.80	3.8	1.8	–	2.0
Compressive strength after 1500 in.-lb/in. impact, ksi	22	42	28	37	34	43

[1]All materials use C6000 ST fiber.

As increases in matrix toughness are made, there will be some sacrifice in hot/wet performance, which can limit composite improvements; therefore, there is a need to consider what can be done from a materials science aspect to improve performance.

An examination of the cross section of an impacted specimen demonstrates that, as can be seen in Figure 1.11, the impact causes considerable splitting and delamination in first generation epoxy materials. This splitting and interply delamination failure are controlled by the inability of the composite to undergo shear deformation. This delamination is a result of the brittle nature of the resins and the laminate construction; that is, the thin resin layers that form between plies are constrained against large shear deformation. Even with the improved matrix materials previously described, there is a limited amount of shear deformation that can be built into the system before the hot/wet deterioration becomes a controlling factor.

One solution to this problem, as shown in Figure 1.20, is to engineer a composite within a composite [31], that is, to take the standard prepreg con-

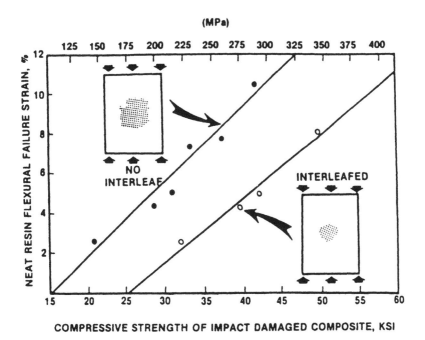

FIGURE 1.20. Interleafing effectively improves the impact resistance of brittle and tough matrix resin systems.

taining 60% by volume fiber with the improved matrix resin and add to it a discreet layer of very high toughness, very high shear strain resin comparable to the matrix of CYCOM®907. The key to this concept is the development of the interleaf material, which can co-cure with the matrix resin and have the flow control to remain as a discreet layer throughout the entire process.

The advantage of this toughened interlayer is shown in Figure 1.21. The compressive strength after impact of the systems shown in Figure 1.18 were increased without modifying the matrix resin. As can be seen, all materials from first generation type through the CYCOM®1806 toughened systems benefit through the addition of an added interlayer. This interlayer functions to suppress initial impact damage, that is, to increase impact resistance. Smaller damage sizes are seen in C-scans of interleafed panels compared to noninterleafed panels impacted at the same energy level. Interleafing is shown to be most effective in improving toughness when combined with a tougher matrix system.

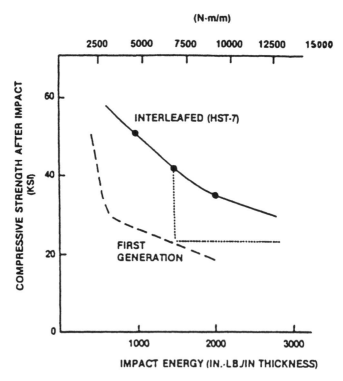

FIGURE 1.21. Comparison of impact resistance of first generation epoxy laminates and interleafed toughened epoxy laminates.

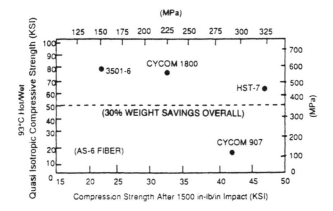

FIGURE 1.22. Trade-off in hot/wet performance and impact resistance for interleafed and single resin composite systems.

A further demonstration of the effect of interlayering on impact damage resistance is given in Figure 1.21. In this figure the residual compressive strength after impact (CASI) is plotted as a function of impact energy for an interleafed material, CYCOM®HST-7, and a first generation epoxy material. Comparing data at 680 kg-m/m (1500 in.-lb/in.), the HST-7 interleafed material has doubled the CAI of first generation materials. From another viewpoint, it is possible to impact the interleafed material at an energy level sufficient to drive the plunger through the structure and still have the same load-carrying capability as the first generation material, which has sustained only barely visible damage.

In addition to toughness, the material's structural performance also is considered. In Figure 1.22 the residual compressive strength after impact is plotted versus the quasi-isotropic laminate compressive strength at 90°C for CYCOM®HST-7 and several other systems. The interleafed system has a residual compressive strength after impact of 320 ± 34 MPa (46.4 ksi) (average for ten panels) and a hot/wet compressive strength of 422 ± 25 MPa (61.2 ± 6 ksi). As the figure demonstrates, the interlayered laminate structural performance ensures a weight savings of more than 30% over aluminum and has greatly improved damage tolerance.

A micromechanics model has been adapted to approximate the effect of interlayering on the laminate compressive strength at elevated temperature. This two-dimensional model was first developed by Rosen [33] and later modified by Greszczuk [34] to predict the compressive strength of unidirectional lamina. The Rosen [33] analysis modeled unidirectional graphite/epoxy as a series of alternating layers of reinforcement (graphite) and resin. The model is directly applicable to an interlayered laminate; the layers of reinforcement are new graphite/epoxy lamina and the toughened interlayer is the resin layer.

The analysis, as modified by Greszczuk, allows for two modes of failure: compressive failure of the individual reinforcing layers and microbuckling of the reinforcing layers. Failure analysis at Cyanamid indicates that these failure modes are related to the test temperature. Compressive failure is found to occur at room temperature and at intermediate temperatures. Microbuckling, on the other hand, occurs at elevated temperatures. The exact values of these temperatures is dependent on the material interleaf and matrix resin systems' T_g.

Greszczuk stated that the compressive failure mode can be described by

$$\sigma_{cu} = \sigma_R \left[V_R + \frac{E_M}{E_R} (1 - V_R) \right] \tag{7}$$

where

σ_R = ultimate compressive strength of reinforcement
E_M = Young's modulus of the interleaf
E_R = axial stiffness of reinforcement
V_R = volume fraction of reinforcement

The equation for σ_{cu} is basically a rule-of-mixtures formulation modified by a factor that includes the relative stiffness of the matrix resin and the reinforcement. It indicates that compressive strength is directly proportional to the volume fraction of reinforcement.

The microbuckling failure modes are predicted by

$$\sigma_{cb} = \frac{G_M}{1 - V_R} + \frac{\pi^2}{3} \left(\frac{h}{L} \right)^2 E_R V_R \tag{8}$$

where

G_M = shear modulus of the interleaf
h = thickness of reinforcement
L = specimen gage length
E_R = axial stiffness of reinforcement
V_R = volume fraction of reinforcement

This equation indicates that the microbuckling failure load (σ_{cb}) is directly proportional to the interleaf shear modulus and to the volume fraction of reinforcement.

Several of the terms included in the equations listed above (G_M, E_M, E_R, and σ_R) are functions of temperature. The effects of temperature on interleaf shear modulus and Young's modulus are determined using a Rheometrics mechanical spectrometer. A second dynamic mechanical spectrometer, a Dyna-Stat, is used to measure the change in reinforcement axial modulus with temperature. The reduction in compressive strength of the reinforcement (that is, the graphite/epoxy layers) with increasing temperature is experimentally determined through a series of elevated temperature compressive tests on noninterleafed lamina specimens.

Figure 1.23 plots the experimental compressive strength of unidirectional laminates versus test temperature. Theoretical results based on the equations discussed previously are also shown in the figure. As the figure illustrates, when the material fails through microbuckling, the compressive strength is greatly reduced over a narrow temperature band. Microbuckling or the onset of microbuckling defines the material's upper use temperature. This temperature may be increased by increasing the interlayer's shear modulus and, more importantly, by retaining that shear modulus at elevated temperatures.

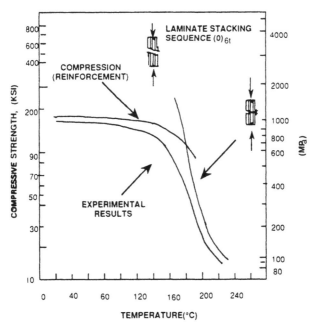

FIGURE 1.23. Analytical model modified to predict interleafed laminate compressive failure modes.

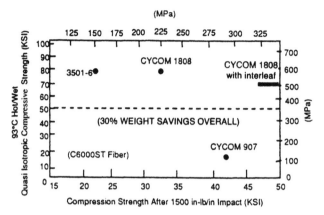

FIGURE 1.24. Structural performance and impact resistance of advanced interleaf system.

The keys to interleaf laminate performance, as predicted by this general model, are to minimize interlayer thickness and to interlayer with resin systems that will retain their shear moduli at elevated temperatures. Applying these criteria, along with the previously stated high-shear-strain-to-failure interleaf selection guideline, has led to the evaluation of several advanced interlayer systems. The total resin content of these systems (interlayer plus matrix resin) is 40% by weight [25]. Preliminary evaluations of these material systems indicate that both structural performance and damage resistance are improved through their application. Figure 1.24 summarizes the 93°C hot/wet compressive strength and residual compressive strength after impact performances of these interlayered systems. Specific individual residual compressive strength data for these systems are presented in Table 1.6. Although it is difficult to compare CSAI data developed at different laboratories as a result of subtle test variables such as compliance of the test fixtures and energy absorbed by the test fixtures, note that

TABLE 1.6. Compressive Strength after Impact of CYCOM1808.[1]

	Compressive Strength after Impact, ksi		
		Interleaf	
Material System	Baseline	S12785-105	S15031-116
CYCM 1808/C6000ST	33	51	55
CYCOM 1808/IM-6	30	43	53

[1]All specimens impacted at 1500 in.-lb/in. thickness.

TABLE 1.7. *Physical Properties of CYCOM 1808/IM6 with Second Generation Interleaf (preliminary data).*

Property	CYCOM 1808/IM6	CYCOM 1808/IM6 (with interlayer)
Unidirectional Properties		
Tensile strength, ksi	325	295
Tensile modulus, Msi	22.4	20.6
Compressive strength, ksi		
23°C/dry	200	180
93°C/dry	184	160
132°C/dry	173	152
93°C/wet	132	143
132°C/wet	69	94
Compressive modulus		
23°C/dry, Msi	22.0	20.0
93°C/dry, Msi	21.5	–
Quasi-Isotropic Properties		
Compressive strength, ksi		
23°C/dry	81	95
93°C/dry	89	–
132°C/dry	85	62
93°C/wet	78	63
132°C/wet	47	50
Compressive strength after impact		
1500 in.-lb/in., ksi	30	53

the CSAI results given in Table 1.5 are comparable to values given for AS-4/poly(etheretherketone) (PEEK) impacted at 600 kg-m/m (1500 in.-lb/in.) [36]. Material property data for CYCOM®1808/IM6 with and without interleaf are presented in Table 1.7. These data fully define the effect of interlayering on structural performance and damage tolerance.

5.4 Damage Tolerance versus Durability

Within the context of aircraft design, MIL-STD-1530A [37] sets forth definitions for key works related to airplane structural integrity: two of these being durability and damage tolerance. In order to evolve a definition of damage tolerance, it is necessary to state definitions for both durability and damage tolerance:

- *Durability* is the ability of the airframe to resist cracking, corrosion, thermal degradation, delamination, wear, and the effects of foreign object damage for a specified period of time.

- *Damage tolerance* is the ability of the airframe to resist failure due to the presence of flaws, cracks, or other damage for a specified period of time.

The above terms have also been considered by industry and cited as

- *Durability* is the amount of abuse or energy a structure or material can absorb without resulting in damage.
- *Damage tolerance* is the load-carrying capability once the material is damaged.

5.5 Materials, Sources of Damage, and Key Considerations for Damage Tolerance

Figure 1.25 focuses attention on damage tolerance issues related to the various classes of material types for which damage tolerance is an issue. Since the focus of this book is on the damage tolerance of fiber-reinforced laminate composites and since the principal commercial composite types in use are PMCs, emphasis will be directed towards this material class. A schematic representation of damage tolerance for PMCs is shown in Figure 1.25. Key elements of damage tolerance in PMCs are shown in Figure 1.26. Regarding the sources of damage, each of the sources indicated is depicted or summarized in Table 1.8.

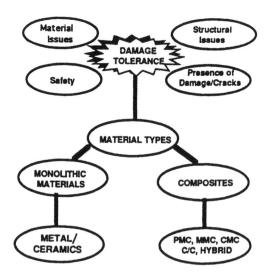

FIGURE 1.25. Damage tolerance schematic.

FIGURE 1.26. Key elements in damage tolerance of PMCs.

These damage sources lead to one of the most significant types of problems associated with continuous fiber polymeric composites (PMCs): their inherent tendency to delaminate. This problem represents the most prevalent type of life limiting failure of advanced composites.

There are various sources of delamination. Delamination can occur by

- interlaminar defects
- out of plane loads
- impact damage

Interlaminar defects may include

- hollow fibers
- delaminations
- fiber breaks
- ply gaps
- excessive porosity, voids
- resin-rich and resin-starved areas

TABLE 1.8. Sources of Damage.

Fabrication/Processing

Fabrication/processing damage (defects) generally include
- Abrasions, scratches, dents, punctures
- Cut fibers
- Knots, kinks
- Improper slicing
- Voids (due to poor processing, highly advanced resin, excess humidity/moisture)
- Resin rich, resin lean areas (improper tensioning)
- Subquality materials

- Cure problems (uncured resin)
- Inclusions, bugs, foreign contamination, etc.
- Tool installation and removal during processing
- Mandrel removal problems (handling)
- Machining problems
- Shipping to propellant processing
- Tool drop (impact damage)
- Proof testing (crazing)

In-Field/Service Problems

In-field/service problems generally include
- Vibration
- Shock
- Lightning damage
- Environment cycling (temperature and humidity)
- Flight loads (fatigue)
- Improper repair (maintenance)

- In-storage creep or handling loads
- Pebble impact (tool drop)
- Scratches, dents, punctures
- Corrosion
- Erosion, dust, sand
- Bacterial degradation

Typical Defects in Composites

- Debonds
- Delaminations
- Inclusions (bugs, foreign contamination, etc)
- Voids, blisters
- Impact damage (tool drop, pebble impact)
- Fiber misalignment
- Cut or broken fibers (dents)

- Abrasions, scratches
- Wrinkles
- Resin cracks, crazing
- Density variations
- Improper cure (soft resin)
- Matching problems (improper hole size, etc.)

- fiber waviness, wrinkles, miscollimation
- foreign particles, contamination, inclusions
- incomplete and/or variable cure
- wrong stacking sequence
- dents, tool impressions, scratches

These interlaminar defects can lead to intense stresses at joints, as shown in Figure 1.27.

The area that represents considerable concern for many of the commercially used continuous fiber–advanced composites is their response under low velocity impact. The damage may not be visible, however, internal delamination can be quite large, resulting in the loss of compression strength and structural integrity. The compressive residual strength of the composite structure can be controlled by the size and location of the delamination. Hence, the design of the composite structure should be made resistant to both formation and growth of delaminations. Since it is impossible to control the low energy impacts that can initiate delaminations, the composite structural designer must assume that such delaminations exist in the laminate a priori and control their growth through intelligent design.

A global view of impact damage in composites is shown schematically in Figure 1.28, while the effect of impact damage in composites is depicted in Figure 1.29.

There are three important issues that can be identified with regard to impact damage in composites:

- type of damage
- damage shape and size
- damage state

Other effects that play an important role in the resultant impact damage in

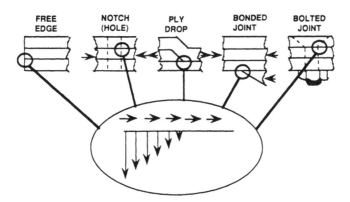

FIGURE 1.27. Interlaminar stresses at joints.

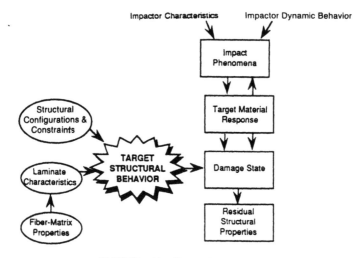

FIGURE 1.28. Global view of impact.

composites is the interactive effect of striker and target. As an example of striker/target interaction and damage development, the effect of a hard striker interacting with a rigid and flexible laminated advanced composite target is shown in Figure 1.30. This example demonstrates that the delamination problem in advanced laminated composite materials is complex in actual structures.

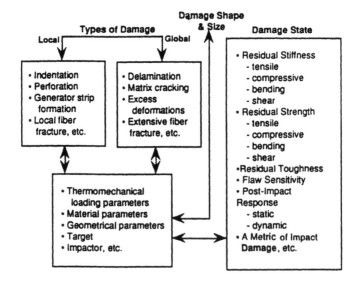

FIGURE 1.29. Impact damage in composites.

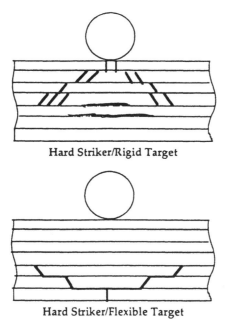

Hard Striker/Rigid Target

Hard Striker/Flexible Target

FIGURE 1.30. Striker/target interaction.

As mentioned previously, the most severe event occurring during service and maintenance is local transverse impact due to the effect of foreign objects. The evaluation of the effects of such impacts, as well as the distribution of the stress waves due to impact as a function of time and space, is highly complex due to the heterogeneous and anisotropic nature of the composite material and is further complicated by loading rate. Hence, loading rate is an important variable to consider.

Another important factor that affects impact events are the boundary conditions of the target. In real situations, boundary conditions may vary from rigidly fixed to free to simply supported, and their exact translation to experimental and analytical models are difficult to determine. Hence, one way to evaluate such variables is by comparing extreme conditions and establishing some empirical factors to adjust the laboratory finding to the real-life structural behavior.

The major failure modes that can occur during loading of composite materials, are fiber fracture, interfiber transverse matrix cracking, and interlaminar fracture or delamination. Fiber fracture rarely occurs in multidirectional laminates under working loads. The presence of a hole may, however, reduce residual strength. Interfiber matrix cracking, which occurs at a low tensile stress transverse to the fiber direction, does not significantly reduce composite performance. On the other hand, delamination is considered to be highly critical to the durability of the composite structure under subsequent long-term (or cyclic) environmental loading service conditions. Even transverse cracking may be crucial under certain loading conditions as a source for additional delamination growth initiating from the transverse crack root. Under flexural loading, two types of delaminations may develop, namely,

- edge delaminations at the tensile zone, close to the outer surface where tensile stresses attain their maximum value
- shear delaminations, close to the neutral plane where interlaminar shear stresses are predominant.

Delaminations may initiate and propagate from free surfaces at holes, cutouts and existing transverse cracks as described before. The chance of predominance of one failure mode above the other depends upon the span to thickness ratio of the beam or panel considered and on the boundary conditions. It is also dependent on loading rate as edge delaminations propagate slowly as a function of tensile axial strain, while shear delamination propagates both quickly and abruptly when the loading energy reaches a critical level. Consequently, a shear delamination mode is expected to be predominant under impact loading, whereas both modes may develop under quasi-static loading.

The current damage tolerance philosophy for composite structures quite justifiably leans heavily on impact damage. It is anticipated that composite structures should sustain life time design loads in a damaged condition up to the condition where a "barely visible" impact damage is detectable. Due to these requirements, the design allowable ultimate strain is limited to ~ 0.4%. However, for further weight savings and for greater damage tolerance performance, design strains of 0.6% are considered highly desirable.

5.6 Damage Tolerance Approaches

Damage tolerance may be achieved in one of two different methods, as shown in Figure 1.31. In fail-safe structures, damage tolerance (and safety) is assured by the allowance of partial structural failure, the ability to detect this failure prior to total loss of the structure, the ability to operate safely with partial failure prior to inspection and the maintenance of specified static strength throughout the service period. Fail-safe structures are usually comprised of multiple elements or load paths such that damage can be safely contained by failing a load path or by arresting a rapidly running crack. In safe crack growth structures, damage tolerance is achieved by sizing the structure, using fracture mechanics or empirical data, such that initial damage will grow at a stable, slow rate and not achieve a size large

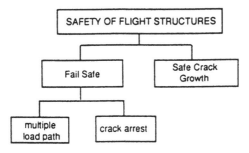

FIGURE 1.31. Safety of Flight Structures

enough to fail the structure prior to detection. A special case within this category is nonpropagating-defect free structures wherein the structure is designed to sufficiently low stress levels for virtually no propagation of the largest defect that would not be detected during a production inspection or that could be incurred in service. Fail safe/damage tolerance is ensured by a specified retained static residual strength and safe crack growth/durability is ensured by damage growth to a specified critical size.

The damage-tolerant requirement, which is illustrated schematically in Figure 1.32, addresses both the residual strength and the damage propagation for the structure under consideration. The residual strength is the amount of static strength available at any time during in-service exposure with damage present. Safety is ensured by designing to requirements wherein damage is never allowed to grow and reduce the residual static strength of the structure below a specified value, such as that corresponding to the maximum load to be experienced in service.

The damage growth curve presents damage size as a function of time.

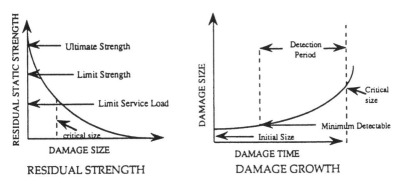

FIGURE 1.32. Damage tolerance requirement.

The period of time required for damage to grow from a minimum detectable size, such as by in-service inspection, to a critical size is the detection period. This critical size generally corresponds to the maximum service load residual strength, although for composites this value could be dictated by stiffness loss at a critical value. If repeated (fatigue) loads are included in the consideration of fail safe/safe crack growth design, then Figure 1.33 can be used.

In order to establish for this discussion the separation of the fatigue aspect of analysis from that of damage tolerance analysis, this chart gives a graphical representation of this separation. Simply stated, damage-tolerant considerations must include the idea of damage detectability. Therefore, this idea acts to provide a real design and application separation. Once detectable, the time to grow to a critical level is the shaded area requiring damage tolerance growth analysis.

An analytical approach to fail safe, safe crack growth can be exemplified

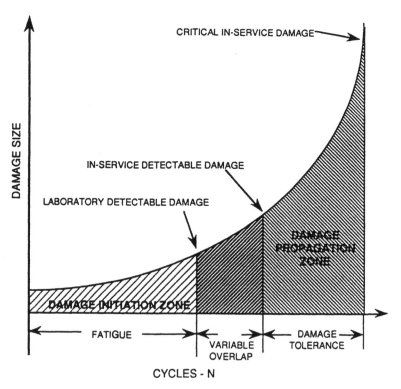

FIGURE 1.33. Cyclie loading effects in damage tolerance design.

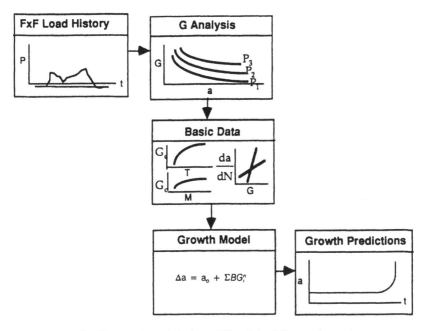

FIGURE 1.34. Aircraft designer DT and durability requirements.

by means of Figure 1.34. Referring to earlier definitions, it would appear that

- Fail safe design relates to damage tolerance.
- Safe crack growth relates to durability.

The methodology, which is to be equally applicable to both durability and damage tolerance analyses, is outlined in Figure 1.34. Its input is a flight-by-flight load history that is coupled to the fracture analysis in order to produce a series of strain-energy release rate/delamination size relationships. The coupon data base can then be entered to judge the potential for static or cyclic growth, including the influence of moisture and temperature. A simple load-by-load growth model produces predictions of delamination size as a function of flight hours. This methodology has many attractive features, but the most important one is that it should be applicable to virtually any laminated composite structure.

With this background, an example (Figure 1.35) has been selected for damage tolerance of composites on some military aircraft. Achievable goals were carefully set for this study by emphasizing manufacturing

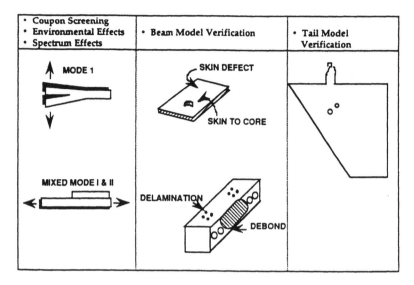

FIGURE 1.35. Example of DT issues for some military aircraft.

defects (such as planar voids) rather than service defects, concentrating on delamination as the major mechanism, developing a fracture–mechanics-based model, and doing it in five increasingly complex phases. The first three phases defined acceptable coupon test specimens, checked the influence of environmental effects, and sought a method for predicting unidirectional delamination growth under spectrum loads. The fourth phase was intended to extend the test base to limited two-dimensional growth in beam specimens representing thin and thick areas of the horizontal tail. These results are not included in this presentation. Correlating delamination growth in a full-scale horizontal tail to that predicted by the coupons was the goal of the final phase. In summary the procedure includes,

- Emphasis on manufacturing defects.
- Concentration on delamination.
- Development of a fracture mechanics-based model.
- Completion of a five-phase program.

Damage tolerance procedures can also be implemented for purposes of making decisions related to structural function as shown in Figure 1.36. In this graphic, defect size versus time are plotted with degree of damage a parameter, as indicated by the labels "mild," "normal," and "severe." The buckling limit of the structural component/configuration is plotted as an upper bound intersecting the damage curves. The effect of damage tolerance flaw size as related to degree of damage is indicated in the graphic.

Composite analysis is also necessary for tracking usage changes to ensure damage tolerance. For Fleet Management activities, it is apparent that the design–usage-related test program is less than adequate for meeting the needs of the Air Force Logistics Command in terms of making maintenance decisions. Tests involving each individual airplane/damage situation are out of the question. For such situations it is concluded that composite analysis is necessary for tracking fleet and individual aircraft usage changes and to make repair decisions (Figure 1.36). Because few "zero-defect" parts are really built, the analysis must be able to relate defect size to service time. Analysis should therefore address the typical flaws occurring in manufacturing (durability flaws), those resulting from unusually bad manufacturing errors (damage tolerance flaws), and larger damage occurring as a result of service usage. The key issue is how the various flaws behave as a function of severe, normal, or limited usage of the airplane. As an example, relationship of flaw size to buckling limit is illustrated in Figure 1.36.

5.7 Overview of the Damage Tolerant Design Approach

Most high-strength graphite-epoxy composite laminates lack ductile behavior at high strains which is characteristic of metals. Composite structures are sensitive to local strain concentrations (such as those that exist in the vicinity of holes and impact damaged regions) resulting in localized failure. This local failure reduces the local stiffness and transfers high

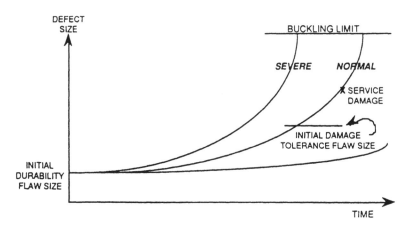

FIGURE 1.36. Composite analysis: tracking damage.

strain concentrations into adjacent regions, which may cause the local damage to propagate. Recent studies indicate that residual strength of impact-damaged laminates can be improved by increasing the fracture toughness of the matrix material, and some investigators anticipate that a high failure-strain fiber could also improve damage tolerance characteristics.

Specific matrix properties that constitute toughness are not well established; however, a high failure strain (greater than 4%) for the matrix with nonlinear stress-strain response at high strains may be a necessary but not a sufficient condition. Resin materials with high fracture toughness (referred to herein as tough resins) may improve the impact damage tolerance characteristics of brittle composites by suppressing the delamination mode of failure. For brittle composites, delamination growth is the dominant failure mechanism observed in tests. The performance of tough-resin composites under cyclic loading is not yet well established. Local shear crippling appears to be the dominant failure mechanism for tough-resin laminates with damage or discontinuities such as holes.

Achievement of damage tolerance for heavily loaded composite aircraft structures can be addressed at several levels of structural complexity. At the material level, tough resins and high failure-strain fibers are being developed. Improved damage tolerance can also be achieved through the use of innovative structural configurations designed to arrest and contain the damage within a region of acceptable size. Recently, there have been advancements in damage containment for both tension- and compression-loaded applications. Finally, damage tolerance can be addressed by the redistributed loads following local, regional failure. In summary, a combination of advancements, including material improvements, damage containment configuration development, and global-local redistribution design methods, are needed to meet the stringent damage tolerance requirements typical of heavily loaded aircraft components such as a wing.

Engineering design considerations must take into account the fact that the composite structure may consist of preformed fabrication flaws and damages due to accidental lateral loads, which will affect the design allowables. Interlaminar flaws are not sensitive to tensile in-plane loading. On the other hand, under compressive in-plane loading, such damage (flaws) are likely to propagate, and consequently, residual compressive strength of damaged laminates is expected to be significantly lower than its virgin reference. Similarly, existing delaminations are expected to propagate under cyclic loading, and hence, fatigue life is liable to be reduced, as compared to that of a reference laminate. In most damage tolerance investigations, residual compressive properties of a laminated panel after impact is plotted versus initial impact energy. (From such an interpretation of data, failure mechanisms and delamination growth can not be revealed and analyzed as a function of loading variables.)

Damage tolerance methodology for advanced composites is not fully mature, primarily due to the complexity in progressive failure modes encountered in service. Except for the case of dominant cracks such as delamination, damage tolerance methodology for interacting failure modes is still continuing to evolve. Current approaches are generally semi-empirical in nature, with considerable bias toward experimental data. One current approach is to assess residual properties and relate them to the damage size and propagation modes, which in turn can be related to the state and distribution of stresses within the laminate. This approach is illustrated in Figure 1.37. Conceptually the interrelationship of crack size, resid

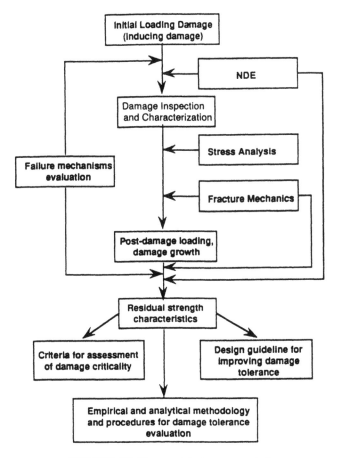

FIGURE 1.37. Damage tolerance procedure.

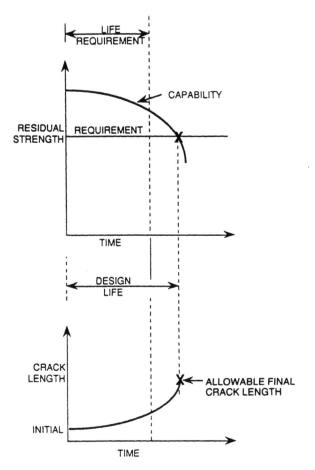

FIGURE 1.38. Interrelationship of residual strength, crack size, and life.

ual strength, and life in the context of achieving damage tolerance is demonstrated in Figure 1.38.

6. REFERENCES

1. Reddick, H. K., Jr. 1978. in "Failure Analysis and Mechanics of Failure of Fibrous Composite Structures," NASA Conference Publication 2278, pp. 129–151.

2. O'Brien, T. K. 1988. "Towards a Damage Tolerance Philosophy for Composite Materials and Structures," NASA TM 100548 (March).

3. Artley, M. G. 1989. "Probabilistic Damage Tolerance Method of Metallic Aerospace Structures," WRWC-TR-3093 (Sept.).

4. Challenger, K. D. 1986. "The Damage Tolerance of Carbon-Fiber Reinforced Composites," A workshop summary, *Composite Structures*, 6:295–318.

5. Baker, A. A., R. Jones and R. J. Callinan. 1985. "Damage Tolerance of Graphite/Epoxy Composites," *Composite Structures*, 4:15–44.

6. Reddick, H. K. Jr. 1978. in "Failure Analysis and Mechanics of Failure of Fibrous Composite Structure," NASA Conference Publication 2278, pp. 129–151.

7. McCarty, J. E. 1982. in "Failure Analysis and Mechanisms of Failure of Fibrous Composite Structures," NASA Conference Publication 2278, pp. 6–77.

8. Wilkins, D. J. 1982. in "Failure Analysis and Mechanisms of Failure of Fibrous Composite Structures," NASA Conference Publication 2278, pp. 67–94.

9. Newaz, G. M. 1992. Battelle Report.

10. Sela, N. and O. Ishai. 1989. "Interlaminar Fracture Toughness," *Composites*, 20:123.

11. Demuts, E. 1989. "Damage Tolerance of Composites," in *Proceedings of the American Society for Composites*, VPI and SU, p. 425.

12. MIL-STD-83444.

13. Evans, R. E. and J. E. Masters. 1987. "A New Generation of Epoxy Composites for Primary Structural Applications: Materials and Mechanics," *ASTM STP 937*, pp. 413–436.

14. Chamis, C. C., M. P. Hanson and T. T. Serafini. 1973. "Criteria for Selecting Resin Matrices for Improved Composite Strength," *Modern Plastics* (May).

15. Williams, J. G. and M. D. Rhodes. 1981. "The Effects of Resin on the Impact Damage Tolerance of Graphite-Epoxy Laminates," NASA-TM-83212 (October).

16. Ying, L. 1983. "Role of R = Fiber/Matrix Interphase in Carbon-Fiber Epoxy Composite Impact Toughness," *SAMPE Quarterly*, 14(3):26.

17. Starnes, J. H. and J. G. Williams. 1982. "Failure Characteristics of Graphite-Epoxy Structural Components Loaded in Compression," NASA-TM-84552 (Sept.).

18. Byers, B. A. 1980. "Behavior of Damaged Graphite/Epoxy Laminates under Compression Loading," NASA Contract Report 159293 (August).

19. Ashiazawa, M. 1983. "Improving Damage Tolerance of Laminated Composites Through the Use of New Tough Resins," in *Proceedings of Sixth Conference on Fiberous Composites in Structural Design*, Watertown, MA: Army Report AMMRC MS 83-21, Army Mechanics Research Center (November).

20. British Aerospace. 1981. Aircraft Group, Preston, Lancashire Report 810/PR/38.

21. Hawthorne, H. M. and E. Teghtsoonian. 1975. "Axial Compression Fracture in Carbon Fibers," *Journal of Material Science*, 10(January):41–51.

22. Greszczuk, L. B. 1972. "Microbuckling of Unidirectional Composites," AFML-TR-71-231 (January).

23. Greszczuk, L. B. 1972. "Failure Mechanics of Composites Subjected to Compressive Loading," AFML-TR-72-107 (August).

24. Lager, J. R. and R. R. June. 1969. "Compressive Strength of Boron-Epoxy Composites," *Journal of Composite Materials*, 3(January):48–56.

25. Chamis, C. C. 1969. "Failure Criteria for Filamentary Composites," NASA-TN-D-5367 (August).

26. Argon, A. S. 1972. *Fracture of Composites, Treatise on Material Science and Technology, Vol. 1*, London: Academic Press Inc.

27. Ewins, P. D. 1974. "Tensile and Compressive Test Specimens for Unidirectional Carbon Fibre Reinforced Plastic," RAE Report 72237.

28. Sednar, G. and R. K. Watterson. "Low Cycle Compressive Fatigue Failure of E Glass-Epoxy Composites," ASRL-TR-162-2, Cambridge, MA: M.I.T. Aeroelastic and Structures Research Lab.

29. Mazzio, V. P., R. L. Mehan and J. V. Mullins. 1973. "Basic Failure Mechanisms in Advanced Composites Composed of Epoxy Resins Reinforced with Carbon Fibres, NASA-CR-134525 (June).

30. Dorey, G. 1980. "Relationships between Impact Resistance and Fracture Toughness in Advanced Composite Materials," in *Proceedings of AGARD Conference, Effect of Service Environment on Composite Materials,* AGARD-CP-288, pp. 1-9.

31. Hirschbuehler, L. R. 1980. "A Comparison of Several Mechanical Tests Used to Evaluate the Toughness of Composites," AGARD-CP-288, pp. 61-73.

32. Krieger, R. B. 1984. "The Relation between Graphite Composite Toughness and Matrix Shear Stress-Strain Properties," in *Proceedings of the 29th National SAMPE Symposium, Technology Vectors,* Covina, CA: Society for the Advancement of Materials and Process Engineering, pp. 1570-1584.

33. Rosen, B. W. 1965. *Fiber Composite Materials,* Metals Park, OH: American Society of Metals, Chap. 3.

34. Greszczuk, L. B. 1974. "Microbuckling of Lamina-Reinforced Composites," in *Composite Materials: Testing and Design* (Third Conference), *ASTM STP 546,* Philadelphia: American Society for Testing and Materials, pp. 5-29.

35. Hirschbuehler, L. R. 1985. "An Improved 270°F Performance Interleaf System Having Extremely High Impact Resistance," in *Proceedings of the 30th National SAMPE Symposium, Advancing Technology in Materials and Processes,* pp. 1335-1346.

36. Wilson, R. D. 1984. "Composite Wing Panel Durability and Damage Tolerance Technology Development," ACEE Composite Structures Technology, NASA Contractor Report 172358, p. 58.

37. MIL-STD-1530A.

CHAPTER 2

Analytical Methodology

1. INTRODUCTION

The opening chapter served to introduce the concept of damage tolerance, sources of damage, and damage tolerance drivers. Among the sources of damage, impact-induced damage for many composite applications is considered as the primary damage concern. Associated with this type of damage event, it is important to establish damage measures, some of which have been suggested for composites to include [1]:

- four-inch long, two-hundredths–inch deep surface scratch
- two-inch diameter equivalent delamination
- one-tenth-inch indentation
- 100 ft-lbs of delivered energy

In pursuing analytical developments associated with impact type events, it is important to understand the physical phenomena involved with such events, which include

- Structural dynamic response
- Hertzian contact effects
- In-plane and through-the-thickness wave effects

These effects are illustrated in Figure 2.1.

The relative importance of the effects shown are dependent upon the inherent characteristics of the striker and the target, as well as the interactive effects between the two. For example, the governing factors needed to define the event over the loading range extending from quasi-static to dynamic are not clearly understood. This can be demonstrated by the following remarks, which relate to striker mass, velocity, and target contact stiffness. Impact

53

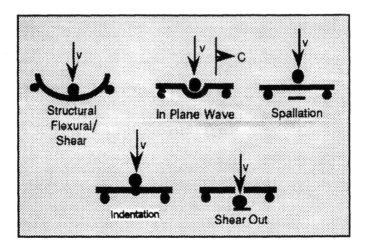

FIGURE 2.1. Schematic of impact events [2].

force is found to be high for the striker/target characteristics found in Table 2.1.

The effect of the boundary conditions may also play a role, as illustrated in Table 2.2.

Thus, measures for classifying what constitutes an impact event in terms of striker/target interactions are necessary in order to quantify the system response. It has been suggested, for example, that the type of impact response may not be driven by the mass, nor by the velocity, but by other quantifiers. Some suggested measures are

(I)	Pulse Period/Target Period	Response
	$< 1/4$	Impact
	$1/4 < t/T < 4$	Vibratory
	> 4	Quasi-static
(II)	Striker Frequency/Target Frequency	Response
	$\ll 1$	Quasi-static
	$\gg 1$	Dynamic

In addition to the structural response measures, a classification of the striker relative to the target as being flexible versus rigid and a geometrical classification of the target as being thick or thin are also necessary. For instance, regarding striker classification, a parameter based upon normal im-

TABLE 2.1. Striker/Target Interaction.

Striker/Target Characteristics	Low	High	Low	High
			Impact force	
Striker velocity		x		x
Striker mass	x			x
Target contact stiffness		x		x

pact of a striker impinging upon a rigid target has been proposed in Reference [3] as

$$\beta = \frac{2(v_y - v_a)^2 c_o}{(v - v_a)}$$

where

v = striker velocity
v_a = target velocity
$v_y = v_a + \sigma_y / c_o \varrho$
c_o = the elastic compressive wave speed in the striker
σ_y = the yield strength of the striker
ϱ = the specific mass of the striker

The striker classification has been denoted by

(III)	Class	Flexible	Semi-Rigid	Rigid
	β	<0.1	$0.1 < \beta < 1.0$	$\beta > 1.0$

In the above classification, the effect of target rigidity can be incorporated through the term v_a.

TABLE 2.2. Striker/Target Boundary Condition Effects.

Boundary Conditions	Low	High
Less important		
Striker velocity	x	
Striker mass		x
More important		
Striker velocity		x
Striker mass	x	

Geometrical classification of targets has also been proposed [4] as

(IV)   • semi-infinite
    • thick
    • intermediate
    • thin

A dimensionless index number has been introduced, expressed by

$$n = \frac{C_t}{C_s} \cdot \frac{(L/d)}{(h/d)}$$

where

C_t = the elastic wave speed in the target
C_s = the elastic wave speed in the striker
(L/d) = the ratio of the striker length to the diameter
(h/d) = the ratio of the target thickness to the striker diameter

A target classification can thus be established as

thick = $n < 1$
intermediate = $1 < n < 5$
thin = $n > 5$

Although each of these measures has merit, none of them, by themselves, represents a definitive quantifier. For example, measure (I) can give a sense for classifying the event; however, it does not provide information on the striker and target interaction. Measures (III) and (IV) provide information on the striker/target material interaction, but they do not provide a guide as to a velocity regime for analyzing target damage. The structural response measure (II) provides a sense to the velocity regime of the striker relative to the target; however, the response is broadly bifurcated as quasi-static and dynamic. The notation of an intermediate velocity regime with a corresponding dynamic response is not identified. All of these factors raise questions as to how to achieve appropriate quantifiers for classifying the impact event, which, in turn, can be useful in developing analytical measures for damage, damage growth, and subsequent failure. It is necessary to identify those elements and parameters associated with the impact event, which are needed for developing a classification schedule. Among the quantifiers needed are such definitions as the velocity regime, that is,

• low velocity

- intermediate velocity
- high velocity

striker, target interaction, for example,

- rigid target/flexible striker
- rigid striker/flexible target

and what parameters control the impact event,

Striker	Target
mass	mass
material	material
velocity	geometry
shape	thickness
size	ply lay-up
angle of incidence	

These factors, as well as others, are needed to quantify the impact event. Models developed for analyzing the impact event may not address all of these elements. Attempts have been made in the representative analytical models cited to provide a table of striker/target information based upon the preceding discussion. In addition, Figure 2.2 describing the impact event is included as a useful guide to describe the event.

A number of analytical developments have been described in the literature, all of which can be broadly classified as being

- empirical
- semi-empirical
- numerical
- analytical

The approaches listed can also be divided, in the broadest sense, into two classifications related to accompanying damage:

- *Local damage* or near field damage at and under the point of impact. This damage is governed by the impact force and results in target penetration/perforation.
- *Global damage* or far field damage that occurs as a result of the impact but is governed by bending and wave propagation effects.

Generally speaking, the damage induced both locally and globally can be described in terms of the following mechanisms:

- matrix cracking

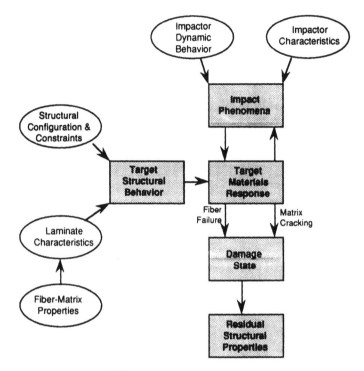

FIGURE 2.2. Impact event flow chart.

- interfacial debonding
- crack coupling
- delamination
- fiber fracture
- indentation
- composite failure

Thus, the study of impact problems and associated analytical models has been divided into three broad classifications;

- deformation mechanics
- damage mechanics
- residual strength degradation

Each of the models cited in the following paragraphs has been related to one of the above-mentioned classifications. In turn, the information generated from these analyses has been used to develop predictions of the

- damage region

- damage growth
- damage effects

For each of these models, basic information as well as a tabulation of important parameters for reference to the reader and for model comparison is presented.

MODEL I

Reference:	Greszczuk [5–7]
Model classification:	deformation mechanics
Impact velocity regime:	low
Striker/target characteristics:	rigid/flexible

EXPERIMENTAL PARAMETERS

Striker	Target
mass – small	mass – large, small
material – steel	material – B/Ep, El/Ep, Gr/Ep
geometry – spherical	geometry – half space
striker incidence – normal	thickness – half space
velocity – low	ply layers – unidirectional, pseudo-isotropic
	boundary conditions – half space

---

2. ANALYSIS—MODEL I

2.1 Isotropic Materials

The model developed is based upon the impact response of isotropic materials with essential features of the model described as follows. The model consists of a spherical striker impacting a half space with the analytical solution based upon combining the dynamic solution for impact of bodies with that of the static solution for pressure between two bodies in contact.

The mass and velocity of the target materials are given by m_t and v_t, respectively, and the striker mass and velocity by m_s and v_s. Using Newtonian mechanics as shown in Figure 2.3, one can write

$$m_t \frac{dv_t}{dt} = -P$$

$$m_s \frac{dv_s}{dt} = -P \tag{1}$$

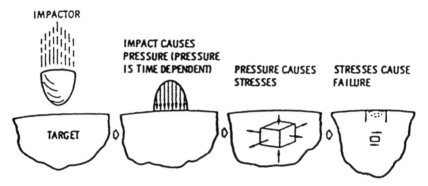

FIGURE 2.3. Sequence of impact event interactions.

The velocity of approach of striker and target at the point of contact can be expressed as

$$\dot{\alpha} = v_t + v_s \tag{2}$$

For the case where the contact time between the striker and target is long in relation to the natural periods of the respective elements, a Hertzian contact law established for static conditions can be used to describe the impact event:

$$P = n\alpha^{3/2} \tag{3}$$

The value of α at maximum compression can be written as

$$\alpha_1 = \left(\frac{5v^2}{4nn_1}\right)^{2/5} \tag{4}$$

where v is the velocity of approach of the two elastic bodies ($v = v_t + v_s$) and

$$n = \left[\frac{4}{3\pi(k_t + k_s)}\right]\sqrt{R_s}$$

where R_s is the radius of the striker

$$n_1 = \frac{1}{m_t} + \frac{1}{m_s}$$

$$k_t = \frac{1 - v_t^2}{\pi E_t}$$

$$k_s = \frac{1 - v_s^2}{\pi E_s}$$

For static indentation of a sphere pressed into a flat surface by a force P, the relation between P, α, and the radius of surface contact a can be written as

$$\alpha = \left[\frac{9\pi^2}{16}p^2\frac{(k_t + k_s)}{R_s}\right]^{1/3} \tag{5}$$

$$a = (R_s)^{1/2}\left(\frac{5v^2}{4nn_1}\right)^{1/5} \tag{6}$$

A solution for P and a in terms of the parameters of striker and target can be obtained by using the expression for α_1 and α, thus

$$P = \frac{\sqrt{R_s}}{3\pi(k_t + k_s)}\left(\frac{5v^2}{4nn_1}\right)^{3/5} \tag{7}$$

$$a = (R_s)^{1/2}\left(\frac{5v^2}{4nn_1}\right)^{1/5} \tag{8}$$

The pressure distribution over the contact surface "a" can be expressed as

$$q_{x,y} = q_o\left[1 - \frac{x^2}{a^2} - \frac{y^2}{a^2}\right]^{1/2} \tag{9}$$

and q_o is the surface pressure at the center of contact, that is at (x,y), equal to zero. By adding the pressures acting over the contact area and equating the result to P, we find

$$q_o = \frac{3P}{2\pi a^2} \tag{10}$$

Further, by combining expressions (7), (8), (9), and (10) and considering polar coordinates, the following expressions for q_o and a are obtained:

$$q_o = \frac{2E_tE_s}{\pi\sqrt{R_s}[(1 - v_t^2)E_s + (1 - v_s^2)E_t]}$$

$$\times \left\{\frac{15[(1 - v_t^2)E_s + (1 - v_s^2)E_tm_sm_tv^2]}{16\sqrt{R_s}E_tE_s(m_s + m_t)}\right\}^{1/5}(1 - r^2/a^2)^{1/2} \tag{11}$$

$$a = R_s^{1/2}v^{2/5}\left\{\frac{15m_sm_t[E_s(1 - v_t^2) + E_t(1 - v_s^2)]}{16\sqrt{R_s}(m_s + m_t)E_sE_t}\right\}^{1/5} \tag{12}$$

In the above expressions, r is the radial distance from the center of the area of contact. Thus, for a given approach velocity v, and using Equations (11) and (12), the maximum pressure distribution over the contact area can be found. With the above expressions, it is possible to obtain

- the maximum pressure occurs at,

$$q_o \text{ (max)} @ \frac{t_o}{2} \tag{13}$$

- the impact duration time

$$t_o = 2.94 \frac{\alpha}{v} \tag{14}$$

A typical indentation time graphic for the impact event is shown in Figure 2.4.

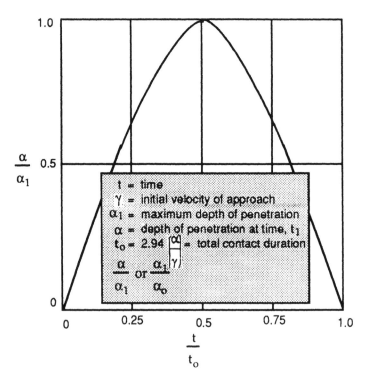

FIGURE 2.4. Striker indentation vs. dimensionless time.

Once the force-time distribution and contact area are established as a function of impact velocity, the internal stress distribution can be found using FEM computer codes. Codes used for this purpose include SAAS III, ASAAS, NASTRAN, ANSYS, MARC, SAP, and NISA.

The maximum tensile, compressive, and shear stresses (σ_t, σ_c, and σ_s) in semi-infinite isotropic solids can be found in terms of the surface pressure using the following equations:

$$J_t = \left(\frac{1 - 2v_t}{3}\right) q_i$$

$$\sigma_c = q_i \tag{15}$$

$$\sigma_s = \left[\frac{1 - 2v_t}{4} + \frac{\sqrt{2(1 + v_t)^3}}{9}\right] q_i$$

where q_i is the maximum pressure at a given time t_i

2.2 Composite Materials

An extension of the previous results for impacted isotropic, pseudo-isotropic solids, to the case of semi-infinite and flexible composite targets, has also been reported by Greszczuk. The particular composite material configurations studied are based upon the codes available for evaluating the triaxial stress state of the impacted medium. The specific composite material configurations studied are

- pseudo-isotropic materials
- orthotropic materials

In the study of these materials, examples illustrating how the impact response and failure modes are influenced by selected parameters has been reported. Included in these parameter studies are

- fiber and matrix properties
- fiber orientation
- stacking sequence
- target thickness
- target curvature

A summary examining the important parameters cited follows.

2.3 Summary of Model I

- *Fiber:* damage resistance increases as fiber strength increases and fiber modulus decreases.
- *Matrix:* damage resistance increases as matrix strength increases and matrix modulus decreases.
- *Ply-orientation:* cross-ply construction resists damage more effectively than unidirectional and multidirectional.
- *Ply sequencing:* Ply sequencing through the thickness is more damage resistant than nondispersed ply sequencing.
- *Damage visibility:* internal damage (matrix cracking, delamination) can be severe without evidence of surface damage.
- *Target thickness:* thick targets show indentation while thin target damage occurs on the backface.
- *Target preload:* target preload affects the magnitude of the impact velocity causing an explosive damage mode.
- *Target curvature:* target curvature affects the impact parameters and failure modes.

MODEL II

Reference:	Shivakumar, Elber, Illg [8]
Model classification:	deformation mechanics
Impact velocity regime:	low
Striker/target characteristics:	rigid/flexible

EXPERIMENTAL PARAMETERS

Striker	Target
mass – large	mass – small
material – steel, aluminum	material – Gr/Ep
geometry – spherical	geometry – circular plate
striker incidence – normal	thickness – thick, intermediate, thin
velocity – low	ply layers – quasi-isotropic
	boundary conditions – clamped, simply supported, movable, immovable

---

3. ANALYSIS—MODEL II

Two models are described, the first an energy-balance model used to predict the impact force magnitude, while the second, a spring mass, is used to obtain information on the impact force. Each model is separately described.

3.1 Energy-Balance Model

This model is used to predict the impact force magnitude with key features of the model described below.

The kinetic energy of the striker is equated to the strain energy of the target, including that due to contact, bending, transverse shear, and membrane deformations. It is assumed that the

- Impact duration is greater than the stress wave travel to the boundary.
- Higher response modes are neglected.
- Energy losses from material damping, surface friction are neglected.

A graphic of the impact event, both before and after impact, is shown in Figure 2.5

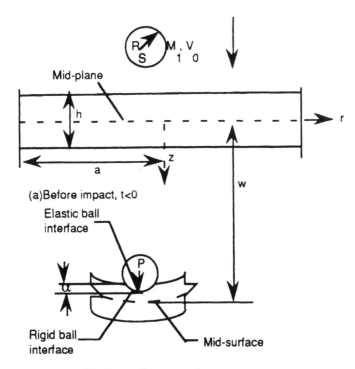

FIGURE 2.5. Schematic of impact event.

3.1.1 IMPACT FORCE DEFORMATIONS

For $t > 0$ these deformations consist of

- contact deformations
- transverse plate deflections

The former relates to the striker and target interaction and is based upon the Hertz contact law, while the latter consists of

- bending denoted by w_b
- shear denoted by w_{sh}
- membrane denoted by w_m

For values of $w/h \leq 2$, membrane effects are neglegible.

3.1.2 TARGET/STRIKER INTERACTION

Target/striker interaction involves two important elements: contact area and plate load. The contact area is dependent on the force and moduli of the striker and target, while the plate load is dependent on the plate bending (K_b), shear (K_s), and membrane stiffness (K_m). The plate bending and membrane stiffnesses as used are derived using variational methods for the case of simply supported and clamped boundary conditions. The shear stiffness K_s is derived, assuming a distributed impact over the contact area.

Thus, the governing equation for the energy balance model is

$$\frac{1}{2}m_s v_o^2 = E_c + E_{bs} + E_m \tag{16}$$

In order to solve the problem, it is necessary to evaluate each term on the right-hand side of Equation (16).

3.1.3 CONTACT ENERGY E_C

The impact force P and contact deformation α are related through the Hertz contact law

$$P = m\alpha^{3/2} \tag{17}$$

with m the contact parameter dependent upon the material and geometrical properties of striker and target. For a spherical isotropic striker and transversely isotropic composite plate, the contact parameter is given by

$$n = \frac{4\sqrt{R_s}}{3\pi(K_1 + K_2)}$$

$$K_1 = \frac{(1 - \nu_s^2)}{\pi E_s}$$

$$K_2 = \frac{\sqrt{A_{22}}\left[\sqrt{A_{11}A_{22}} + (G_{zr}) - A_{12} + G_{zr})^2\right]^{1/2}}{2\pi\sqrt{G_{zr}(A_{11}A_{22} - A_{12}^2)}}$$

$$A_{11} = E_z(1 - \nu_r)\beta$$

$$A_{22} = \frac{E\beta(1 - \nu_{zr}^2)\delta}{(1 + \nu_r)}$$

$$A_{12} = E_r\nu_{zr}\beta$$

$$\beta = \frac{1}{1 - \nu_r - 2\nu_{zr}^2\delta}$$

$$\delta = E_r/E_z$$

where

E_s, ν_s = the modulus and Poisson's ratio of the striker
E, G, ν = the moduli and Poisson's ratio of the target
r, z = the radial and thickness coordinates

Thus, the contact energy is obtained using

$$E_c = \int_o^a P d\alpha \tag{18}$$

Using the Hertz contact law for $P(\alpha)$, the contact energy E_c can be written as

$$E_c = \frac{2}{5}\frac{P^{5/3}}{m^{2/3}} \tag{19}$$

3.1.4 BENDING, SHEAR ENERGY E_{bs}

To determine the target bending, shear energy E_{bs}, it is necessary to relate the target reactive force P_{bs} to the transverse plate deflection. This relation is given by

$$P_{bs} = K_{bs}w \tag{20}$$

where the quantity K_{bs} represents an effective, bending shear stiffness denoted by

$$K_{bs} = \frac{K_b K_s}{K_b + K_s}$$

A menu of bending stiffness values for several edge condition and boundary conditions are available in the references cited. As an example, consider the following case:

- boundary condition – clamped
- edge condition – immovable

The corresponding bending stiffness, K_b, is

$$K_b = \frac{4\pi E_r h^3}{3(1 - \nu_r^2)a^2} \tag{21}$$

The shear stiffness is given by

$$K_s = \frac{4\pi G_{zr}}{3}\left(\frac{E_{rz}}{E_{rz} - 4\nu_{rz}G_{zr}}\right)\left(\frac{1}{4/3 + \log a/a_c}\right) \tag{22}$$

where the contact radius, a_c, is expressed as

$$a_c = \left[\frac{3\pi}{4}P[K_1 + K_2]R_s\right]^{1/3}$$

The bending shear energy can thus be expressed as

$$E_{bs} = \frac{1}{2}K_{bs}w^2 \tag{23}$$

where K_{bs} has been previously defined.

3.1.5 MEMBRANE ENERGY E_m

The target reactive force for membrane effects can be expressed as

$$P_m = K_m w^3 \tag{24}$$

Correspondingly, the membrane energy is expressed as

$$E_m = \frac{1}{2} K_m w^4 \tag{25}$$

The expression for the membrane stiffness, K_m, is dependent upon the edge and boundary conditions selected. For example, for the case of,

- boundary condition — clamped
- edge condition — immovable

the corresponding membrane stiffness is given by

$$K_m = \frac{(353 - 191\nu_r)\pi E_r h}{648(1 - \nu_r)a^2} \tag{26}$$

The complete menu of membrane, as well as bending shear stiffness parameters, is shown in Table 2.3. The energy balance equation can then be expressed as

$$m_s v_o^2 = K_{bs} w^2 + \frac{K_m w^4}{2} + \frac{4}{5}\left[\frac{K_{bs} w + K_m w^3)^5}{n^2}\right]^{1/3} \tag{27}$$

3.1.6 SOLUTION PROCEDURE

The calculation procedures used for evaluating Equation (27) are

- solving for w using numerical schemes
- solving for v_o for a given w

The impact force P is then found using

$$P = K_{bs} w + K_m w^3 \tag{28}$$

3.2 Spring-Mass Model

While the Energy-Balance model provides information on the impact force, information on the force time history through the impact event is not

TABLE 2.3. *Bending and Membrane Stiffness Parameters of Centrally Loaded Plates.*

Boundary Conditions	Edge Conditions	Bending Stiffness	Membrane Stiffness Parameters
Clamped			
	Immovable	$\dfrac{4\pi E_r h^3}{3(1-\nu_r^2)a^2}$	$\dfrac{(353-191\nu_r)\pi E_r h}{648(1-\nu_r)a^2}$
	Movable	$\dfrac{4\pi E_r h^3}{3(1-\nu_r^2)a^2}$	$\dfrac{191\pi E_r h}{648 a^2}$
Simply supported			
	Immovable	$\dfrac{4\pi E_r h}{3(3+\nu_r)(1-\nu_r)a^2}$	$\dfrac{4\pi E_r h}{(3+\nu_r)a^2}\left\{\dfrac{191}{648}(1+\nu_r)^4+\dfrac{41}{27}(1+\nu_r)^3+\dfrac{32}{9}(1+\nu_r)^2\right.$ $+\dfrac{40}{9}(1+\nu_r)+\dfrac{8}{3}+1(1+\nu_r)\dfrac{(1+\nu_r)^4}{4}+2(1+\nu_r)^3$ $+\dfrac{40}{9}(1+\nu_r)+\dfrac{8}{3}+1/(1+\nu_r)\left[\dfrac{(1+\nu_r)^4}{4}+2(1+\nu_r)^3\right.$ $\left.\left.+8(1+\nu_r)^2+16(1+\nu_r)+16\right]\right\}$
	Movable	$\dfrac{4\pi E_r h^3}{3(3+\nu_r)(1-\nu_r)a^2}$	$\dfrac{4\pi E_r h}{a^2(3+\nu_r)}\dfrac{191}{648}(1+\nu_r)^4+\dfrac{41}{27}(1+\nu_r)^3+\dfrac{32}{9}(1+\nu_r)^2$ $+\dfrac{40}{9}(1+\nu_r)+\dfrac{8}{3}\Big]$

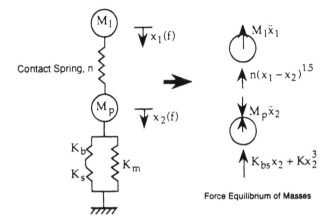

FIGURE 2.6. Spring-mass model.

provided. In order to incorporate this effect, an alternative Spring-Mass Model is introduced. In this model, the striker and target are considered as rigid masses m_s and m_t with the target mass taken as an effective mass. The effective target mass is the target mass plus one-fourth the target mass in order to account for inertial effects. The two masses are connected by springs, as shown in Figure 2.6.

The spring stiffness is represented by a contact spring between the striker and target, with bending, shear, and membrane springs. It should be noted that, for thin targets, that is, for transverse target deflection w less than two-tenths of the target thickness h, small deflection theory can be used. In addition, the assumptions used for the Energy-Balance Model in reference to material damping and surface friction also hold for the Spring-Mass Model.

3.3 Model Development

Using the Hertz contact law, that is,

$$P = n\alpha^{3/2} \tag{29}$$

the contact force between striker and target can be written as (the striker mass m_s is considered to be in contact with the target)

$$P = n(x_1 - x_2)^{1.5} \tag{30}$$

where $w = x_2(t)$ is the transverse deflection of the target and $\alpha = x_1(t) - x_2(t)$ is the contact deformation.

Using Newtonian mechanics for the two degree of freedom model, the problem can be formulated in terms of the governing equations and initial conditions as

Governing Equations

$$m_s \ddot{x}_1 + \lambda n |x_1 - x_2|^{1.5} = 0 \tag{31}$$

$$m_{ts} \ddot{x}_2 + K_{bs} x_2 + K_m x_2^3 - \lambda n |x_1 - x_2|^{1.5} = 0 \tag{32}$$

Note: m_s = striker mass, for $x_t > x_2$; m_t = target mass, for $x_1 < x_2$

Initial Conditions

$$x_1(0) = 0$$

$$x_1(0) = v_o$$

3.4 Solution to Equations

The coupled nonlinear equations are solved using a numerical integration scheme with the solution terminated when the target displacement reaches either zero or a negative value. The impact force is calculated, as in the case of the Energy-Balance Model, using

$$P = K_{bs} w + K_m w^3 \tag{33}$$

with $x(t)$ substituted for w. Thus, the force is obtained.

The two degree of freedom model can be simplified for the case where the striker mass is 3.5 times greater than the total plate mass. The equation of equilibrium can then be expressed as a single degree of freedom system that is

$$m_s \ddot{x}_1 + K_{bs} w + K_m w^3 + \eta \alpha^{1.5} = 0 \tag{34}$$

with a solution generated as in the preceding example. Initial conditions are the same as in the two degree of freedom case.

3.5 Summary of Model II

The Energy-Balance Model has been used to calculate the impact force magnitude for a number of cases. The results have, in turn, been compared with the model introduced by Greszczuk [4–6]. Figure 2.7 shows results

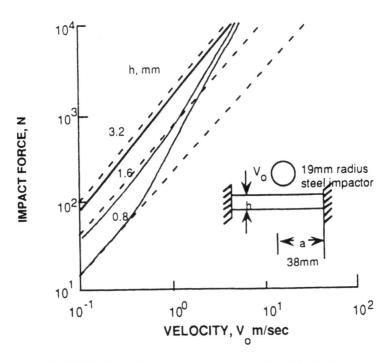

Striker	Steel Sphere
Geometry Striker	19mm
Strike	Hard
Strike	Low
Target	Circular Plate
Geom	38mm
Target	0.8mm, 1.5mm, 3.2mm
Target	Graphite/Epoxy
Target	Quasi-Isotropic
Target	Clamped
Target Edge Conditions	Immovable
Target Hardness	Soft
Target Classification	Thick, Intermediate, Thin

FIGURE 2.7. Impact force vs. velocity for various target thickness.

from one case described in the accompanying table, with results obtained by Greszczuk shown in dotted lines.

Differences between the Energy-Balance (EB) Model and that of Greszczuk (G) indicate that for

- Thicker plates (3.2 mm), the G model predicts higher impact forces than the EB model. This is due to the inclusion of shear effects in the EB model.
- Thin plates (0.8 mm), the inclusion of membrane stiffening in the EB model produces higher impact forces than the G model.

MODEL III

Reference:	Cairnes, Lagace [9]
Model classification:	deformation mechanics
Impact velocity regime:	low
Striker/target characteristics:	rigid/flexible

EXPERIMENTAL PARAMETERS

Striker	Target
mass – small	mass – small
material – steel	material – Gr/Ep
geometry – spherical	geometry – square plate
striker incidence – normal	thickness – thin
velocity – low	ply layers $[(90/0)_2 90]_s$

---

4. ANALYSIS—MODEL III

A global model is developed for rectangular composite plates subjected to impact by a hard spherical striker. Details of the local contact behavior are not considered, but only the nature of the local contact behavior is considered. Due to the inherent anisotropy of composite plates, the normal modes of vibration are coupled, precluding a modal solution. Thus, a solution is generated based upon

- an energy method (Rayleigh-Ritz) used to solve the time varying initial problem
- the local contact behavior which is assumed to be Hertzian

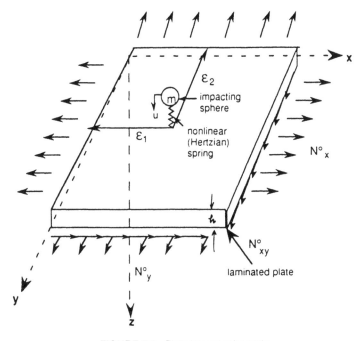

FIGURE 2.8. Plate impact schematic.

The basic plate impact schematic is shown in Figure 2.8. In this figure, the contact spring behavior is assumed to be of the form

$$R = K(\alpha)^{3/2} \tag{35}$$

where R equals the force on the plate, α the local indentation, and K a function of the properties of striker and target.

Other assumptions used in the plate analysis include the following:

- Plate is anisotropic.
- Bending-twisting coupling is included.
- Shearing deformation is considered.

The development of the governing equations for the plate impact problem is based upon using the kinematic and constitutive relations indicated.

4.1 Kinematic Relations

$$\begin{Bmatrix} \chi_x \\ \chi_y \\ \chi_{xy} \end{Bmatrix} = \begin{Bmatrix} \Psi_{x,x} \\ \Psi_{y,y} \\ \Psi_{x,y} + \Psi_{y,x} \end{Bmatrix} \tag{36}$$

and

$$\begin{Bmatrix} \gamma_{xz} \\ \gamma_{yz} \end{Bmatrix} = K \begin{Bmatrix} \Psi_x - \dfrac{\partial w}{\partial x} \\ \Psi_y - \dfrac{\partial w}{\partial y} \end{Bmatrix} \tag{37}$$

where x are the plate curvatures, Ψ are the planar rotations, K is the shearing correction factor (taken as 5/6), and w is the transverse displacement.

4.2 Constitutive Equations

$$\begin{Bmatrix} M_x \\ M_y \\ M_{xy} \end{Bmatrix} = \begin{bmatrix} D_{11} & D_{12} & D_{16} \\ D_{12} & D_{22} & D_{26} \\ D_{16} & D_{26} & D_{66} \end{bmatrix} \begin{Bmatrix} x_x \\ x_y \\ x_{xy} \end{Bmatrix} \tag{38}$$

and

$$\begin{Bmatrix} N_{xz} \\ N_{yz} \end{Bmatrix} = \begin{bmatrix} A_{55} & A_{45} \\ A_{45} & A_{44} \end{bmatrix} \begin{Bmatrix} \gamma_{xz} \\ \gamma_{yz} \end{Bmatrix} \tag{39}$$

4.3 Governing Equations of Equilibrium

Governing equations of equilibrium are derived using minimization of potential and kinetic energy. The expression for the plate bending and shear strain energy are,

Plate Bending Strain Energy

$$U_b = \frac{1}{2} \int_o^a \int_o^b \{x\}^T [D] \{x\} dy \ dx \tag{40}$$

Plate Shear Strain Energy

$$U_s = \frac{1}{2} \int_o^a \int_o^b x \{\gamma\}^T [A] \{\gamma\} dy \ dx \tag{41}$$

The work due to the applied loads consists of both that due to the transverse and in plane loads. The expressions are given by,

Transverse work

$$w_1 = -\int_o^a \int_o^b p(x,y)w(x,y)dy\,dx \tag{42}$$

In-plane loads

$$w_n = \frac{1}{2}\int_o^a \int_o^b \begin{Bmatrix} w_{,x} \\ w_{,y} \end{Bmatrix}^T \begin{bmatrix} N_x^o & N_{xy}^o \\ N_{xy}^o & N_y^o \end{bmatrix} \begin{Bmatrix} w_{,x} \\ w_{,y} \end{Bmatrix} dy\,dx \tag{43}$$

Here,

$p(x,y)$ = the transverse loading on the plate
$w(x,y)$ = the transverse displacement
N_x^o, N_y^o, N_{xy}^o = the average in plane loads

The plate kinetic energy is

$$T_\varrho = \frac{1}{2}\int_0^a \int_0^b \begin{Bmatrix} \dot{\Psi}_x \\ \dot{\Psi}_y \\ \dot{w} \end{Bmatrix}^T \begin{bmatrix} I & 0 & 0 \\ 0 & I & 0 \\ 0 & 0 & P \end{bmatrix} \begin{Bmatrix} \dot{\Psi}_x \\ \dot{\Psi}_y \\ \dot{w} \end{Bmatrix} dy\,dx \tag{44}$$

In the above, a dot refers to a derivative with respect to time, and the transverse inertia matrix and plate rotatory inertia matrix I are defined by

$$[P,I] = \varrho \int_{-h/2}^{+h/2} [1,z^2]dz \tag{45}$$

Here,

h = the plate thickness
ϱ = the mass density of the plate (target)

To solve the problem, the Rayleigh-Ritz method is used, along with assumed displacement mode shapes. The displacements are assumed to be seperable functions of x and y, which leads to the following form for plane rotations and transverse displacements.

$$\begin{aligned}
\Psi_x &= \Sigma A_i(t) f_i(x) g_i(y) \\
\Psi_y &= \Sigma B_i(t) h_i(x) l_i(y) \\
w &= \Sigma c_i(t) m_i(x) n_i(y)
\end{aligned} \tag{46}$$

where A_i, B_i, $C_i(t)$ are time varying modal amplitudes, with i representing the mnth mode.

To solve the problem, beam functions satisfying the displacement boundary conditions are used along with functions for the planar rotations, these being derivatives of the lateral displacements, thus,

$$f_i(x) = m_i'(x)$$

$$h_i(x) = m_i(x)$$

$$g_i(y) = n_i(y) \tag{47}$$

$$l_i(y) = n_i'(y)$$

The prime denotes a spatial derivative.

Using T and V to represent the kinetic and potential energy of the plate, the Lagrangian equations of motion can be developed.

$$\frac{d}{dt}\left[\frac{\partial L}{\partial \dot{x}_i}\right] - \frac{\partial L}{\partial \dot{x}_i} = R_i \tag{48}$$

with the following equations obtained:

$$\begin{bmatrix} I_x & 0 & 0 \\ 0 & I_y & 0 \\ 0 & 0 & M \end{bmatrix}\begin{pmatrix} \ddot{A}_i \\ \ddot{B}_i \\ \ddot{C}_i \end{pmatrix} - \begin{bmatrix} K_{aa} & K_{ab} & K_{ac} \\ K_{ab} & K_{bb} & K_{bc} \\ K_{ac} & K_{bc} & K_{cc} \end{bmatrix}\begin{pmatrix} A_i \\ B_i \\ C_i \end{pmatrix} = \begin{pmatrix} R_{ai} \\ R_{bi} \\ R_{ci} \end{pmatrix} \tag{49}$$

In the above, x_i is the generalized modal amplitude and R_i the modal forces.

The components of the stiffness matrices are found by integrating the assumed mode shapes over the surface of the plate using Simpson's rule. To find the modal forcing amplitude, Figure 2.9, for the impacting mass, is used.

The local indentation is defined as

$$\alpha(t) = u(t) - w(t_1, \xi_1, \xi_2) \tag{50}$$

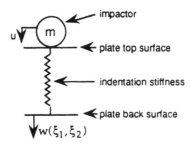

FIGURE 2.9. Impactor target schematic.

where u equals the striker displacement and w equals the plate displacement. The particle equation of equilibrium for the striker is given by,

$$m\ddot{u} = k(\alpha)^{3/2} = R \qquad (51)$$

where m equals the striker mass and R equals the reaction of the striker.

Since loading on the plate is a point loading, the loading may be modelled as a Dirac delta function.

A solution to the system of governing equations can be obtained using simplifications associated with neglect of the rotatory inertia terms. Thus, with neglect of rotatory inertia terms and retaining the shear deformation terms, the governing equations can be written as,

$$[M]\{\ddot{C_i}\} + [K_{cc}^*]\{C_i\} = \{R_{ci}\} \qquad (52)$$

where

$$[K_{cc}^*] = \left[K_{cc} - \begin{Bmatrix} K_{ac} \\ K_{ba} \end{Bmatrix}^T \begin{bmatrix} K_{aa} \ K_{ab} \\ K_{ba} \ K_{bb} \end{bmatrix}^{-1} \begin{Bmatrix} K_{ac} \\ K_{bc} \end{Bmatrix} \right] \qquad (53)$$

To solve these equations, the Newmark implicit integration scheme is used. Also, since the forcing function R is nonlinear during the contact period of striker and target, a predictor corrector approach is used.

4.4 Summary of Model III

The model considered is useful for predicting the dynamic response of plates and for evaluating a number of important parameters associated with the impact event. Some of the important contributions of this model are,

- The mass of the striker has a significant effect on the outcome of the impact event. [Low striker mass, high striker velocity produces high loads.]

- The mass of the target with respect to the striker is equally important. [Low target mass (lower target inertia) produces lower contact forces.]
- Local contact models and global models need to be considered to define the impact problem.
- The impact event can not be defined solely by reference to the initial impactor energy.

MODEL IV

Reference:	Madsen, Morgan, Nuismer [10]
Model classification:	deformation mechanics
Impact velocity regime:	low
Striker/target characteristics:	rigid/flexible

EXPERIMENTAL PARAMETERS

Striker	Target
mass – small	mass – large, small
material – steel	material – Gr/Ep
geometry – hemispherical	geometry – square space
striker incidence – normal	thickness – thin
velocity – low, intermediate	ply layers – [± 18/90₄/ ∓ 18]
	boundary conditions – clamped-free

---

5. ANALYSIS—MODEL IV

A two degree of freedom model consisting of a two-mass, two-spring system is considered using a mass m_1 to model the striker and a mass m_2 to model the target. A schematic of the spring mass configurations used for the analytical model is shown in Figure 2.10.

In the model considered, the contact spring is modelled as linear although a nonlinear representation is considered as more representative.

The equations of motion for the two-mass–two-spring system using Newtonian mechanics can be written as

$$k_1(x_2 - x_1) = m_1 \ddot{x}_1 \tag{54}$$

$$-k_1(x_2 - x_1) - k_2 x_2 = m_2 \ddot{x}_2 \tag{55}$$

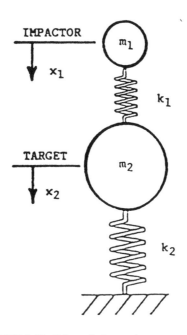

FIGURE 2.10. Schematic two-spring mass model.

where the subscript 1 refers to the striker and the subscript 2 refers to the target.

The initial conditions for the problem are given

$$x_1(o) = x_2(o) = o \tag{56}$$

$$\dot{x}_1(o) = v_1, \dot{x}_2(o) = o \tag{57}$$

A solution to the problem is given as

$$x_1(t) = v_1(A_I \sin \omega_I t + A_{II} \sin \omega_i t) \tag{58}$$

$$x_2(t) = v_1(x_I A_I \sin \omega_I t + X_2 A_{II} \sin \omega_{II} t) \tag{59}$$

The amplitudes are

$$A_I = 1/(1 - X_I/X_{II})\omega_I \tag{60}$$

$$A_{II} = 1/(1 - X_{II}/X_I)\omega_{II} \tag{61}$$

The correspondonding mode shapes are

$$x_I = 1 - (m_1/k_1)\omega_I^2 \tag{62}$$

$$x_{II} = 1 - (m_1/k_1)\omega_{II}^2 \tag{63}$$

with the frequencies ω_I, ω_{II} given by

$$\omega^2 = \frac{(k_1 + k_2)m_1 + k_1 m_2 \mp \sqrt{[(k_1 + k_2)m_1 + k_1 m_2]^2 - 4k_1 k_2 m_1 m_2}}{2m_1 m_2}$$

$$\tag{64}$$

The contact force is determined from

$$F(t) = k_1(x_2 - x_1) = -m_1 v_1(\omega_I^2 A_I \sin \omega_I t + \omega_{II}^2 A_{II} \sin \omega_{II} t) \tag{65}$$

In the present solution, the mode shapes and frequencies are independent of the impact velocity; the impact velocity influences only the magnitude of the displacements and the contact force.

Discussion of the total system response for two special cases is considered.

5.1 Case 1: $\omega_1/\omega_2 \ll 1$

This case is that of large striker mass impact. Since $\omega_1/\omega_2 \ll 1$ this implies that

$$m_1 \gg (k_1/k_2)m_2$$

Also, since the striker mass is large, the corresponding velocity is low and case 1 can be referred to as a low velocity impact event. Expanding the solutions $x_1(t)$ and $x_2(t)$ in powers of ω_1/ω_2 and retaining lower order terms, the impact response can be written as

$$x_1(t) = \sqrt{\frac{m_1 v_1^2}{k_{eff}}} \sin \sqrt{\frac{k_{eff}}{m_1}} t \tag{66}$$

$$x_2(t) = \left(\frac{k_1}{k_1 + k_2}\right) \sqrt{\frac{m_1 v_1^2}{k_{eff}}} \sin \sqrt{\frac{k_{eff}}{m_1}} t \tag{67}$$

Correspondingly, the contact force $F(t)$ is given by

$$F(t) = -\sqrt{k_{eff} m_1 v_1^2} \sin \sqrt{\frac{k_{eff}}{m_1}} t \qquad (68)$$

and

$$k_{eff} = \frac{k_1 k_2}{k_1 + k_2}$$

The impact response indicated is equivalent to Figure 2.11, that is, a system in which m_1 is taken as zero.

The time duration of the impact event is given by

$$t_d = \pi \sqrt{\frac{m_1}{k_{eff}}} \qquad (69)$$

while the time for the fundamental mode to go through one cycle of vibration is of the order of

$$t_c = \sqrt{\frac{m_2}{k_2}} \qquad (70)$$

Since $\omega_1/\omega_2 \ll 1$, this implies that $t \gg t_c$.

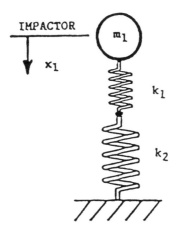

FIGURE 2.11. Schematic two-spring, single-mass model.

This case is equivalent to that previously discussed in the introductory remarks to this chapter, that is, a quasi-static response in which the impact event depends upon the striker/target frequency, ω_1/ω_2, and does not depend upon the impact velocity or the striker/target mass ratio.

5.2 Case 2: $\omega_1/\omega_2 \gg 1$

This case is considered to be a small mass impact event since $\omega_1/\omega_2 \gg 1$. This implies that

$$m_1 \ll \left(\frac{k_1}{k_2}\right) m_2$$

This event can also be referred to as a high velocity impact event. Expanding the solutions $x_1(t)$, $x_2(t)$ in a power series, retaining lower order terms results in the following response equations:

$$x_1(t) = \sqrt{\frac{m_1 v_1^2}{k_2\left(1 + \dfrac{m_2}{m_1}\right)}} \sin \sqrt{\frac{k_2}{m_1 + m_2}}\, t + \sqrt{\frac{m_1 v_1^2}{k_1\left(1 + \dfrac{m_1}{m_2}\right)^3}} \sin \sqrt{\frac{k_1}{m_1} + \frac{k_2}{m}}\, t \tag{71}$$

and

$$x_2(t) = \sqrt{\frac{m_1 v_1^2}{k_2\left(1 + \dfrac{m_2}{m_1}\right)}} \sin \sqrt{\frac{k_2}{m_1 + m_2}}\, t$$

$$- \left(\frac{m_1}{m_2}\right) \sqrt{\frac{m_1 v_1^2}{k_1\left(1 + \dfrac{m_1}{m_2}\right)^3}} \sin \sqrt{\frac{k_1}{m_1} + \frac{k_1}{m_2}}\, t \tag{72}$$

$$F(t) = \sqrt{\frac{k_1 m_1 v_1^2}{1 + \dfrac{m_1}{m_2}}} \sin \sqrt{\frac{k_1}{m_1} + \frac{k_1}{m_2}}\, t \tag{73}$$

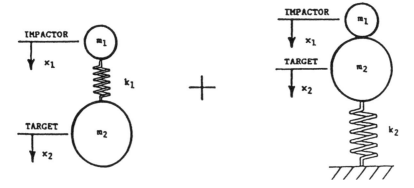

FIGURE 2.12. High and low velocity impact events.

This impact response is equivalent to that of a spring mass system in which the mass m_1 strikes m_2 with k_2 taken as zero. This solution is then superposed upon that of a lumped mass $(m_1 + m_2)$ with k_1 taken as zero and k_2 as nonzero. This model is shown in Figure 2.12.

For this case the time duration of the impact event is given by

$$t_d = \pi \sqrt{\frac{m_1 m_2}{k_1(m_1 + m_2)}} \tag{74}$$

while the cyclic wave reflection for the fundamental mode of structural vibration is given by

$$t_c = \sqrt{\frac{m_2}{k_2}} \tag{75}$$

Since $\omega_1/\omega_2 \gg 1$, it is implied that $t_d \ll t_c$. This response is classified as a dynamic response since higher modes of vibration are important.

5.3 Summary of Model IV

The two-mass–two-spring model described results in a different response for the impact event depending upon the ratio of the striker frequency (striker mass to contact stiffness) to the structural frequency (structural mass–structural stiffness).

Some specific remarks follow:

- For a striker frequency ≪ than the target frequency response, the event is classified as being quasi-static. For such events an energy balance model with a static structural deflection solution describes the impact event and can be used to determine the contact force.
- For a striker frequency ≫ than the target structural frequency, the event is described as dynamic. For this case the higher vibrational modes must be considered in the system response.

Some shortcomings of this model are that,

- Nonlinearities in the contact stiffness are neglected.
- The lumped mass model ignores mass and stiffness distribution.
- The striker-target loss of contact is ignored.

MODEL V

Reference:	Shivakumar, Elber, Illg [11]
Model classification:	damage mechanics
Impact velocity regime:	low
Striker/target characteristics:	rigid/flexible

EXPERIMENTAL PARAMETERS

Striker	Target
mass – large	mass – small
material – steel	material – Gr/Ep
geometry – hemispherical	geometry – circular plate
striker incidence – normal	thickness – thin
velocity – low	ply lay-up – quasi-isotropic
	boundary conditions – clamped

---

6. ANALYSIS—MODEL V

A steel sphere is used to apply loads to a circular plate, such that a low velocity impact event can be modelled as a static equivalent impact load. This is shown in Figure 2.13.

The plate itself is circular in shape, with clamped boundary conditions, and is made of graphite/epoxy with a quasi-isotropic lay-up. A schematic of the laminated plate used as a basis for a numerical analysis is shown in Figure 2.14.

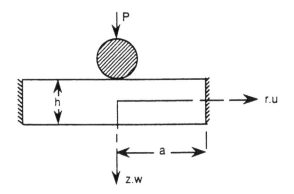

FIGURE 2.13. Impact event diagram.

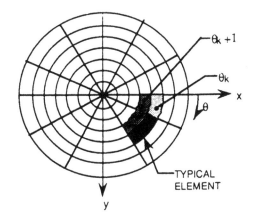

(a) Plan view of an ith layer.

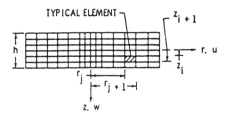

(b) Sectional view at $\theta = \theta_k$.

FIGURE 2.14. Schematic of elements modelling the impact event: (a) plan view of an ith layer; (b) sectional view at $\Theta = \Theta_k$.

The analysis is based upon the principal of minimization of the total potential energy with three key assumptions:

(1) The quasi-isotropic circular plate deforms axisymmetrically under an axisymmetric load.
(2) The deformed shape of the plate for large and small deflections is the same.
(3) The transverse shear deformation is negligible, since $a/h > 10$.

6.1 Displacement Components

The displacement components in the transverse and radial directions are written, respectively, as

$$w = w_o \left[1 - \left(\frac{r}{a}\right)^2 + 2\left(\frac{r}{a}\right)^2 \ln\left(\frac{r}{a}\right) \right] \tag{76}$$

$$u = r(a - r)(C_1 + C_2 r) \tag{77}$$

These displacement functions satisfy the plate boundary conditions at $r = 0$ with $u = dw/dr = 0$. The constants C_1, C_2 are determined by minimizing the total potential energy of the deformed plate.

6.2 Strain-Displacement Relations

The radial, tangential, and shear strains are given, respectively, by

$$\epsilon_r = \frac{du}{dr} + \frac{1}{2}\left[\left(\frac{du}{dr}\right)^2 + \left(\frac{dw}{dr}\right)^2\right] - z\frac{d^2w}{dr^2} \tag{78}$$

$$\epsilon_\theta = \frac{u}{r} - \frac{z\,dw}{r\,dr} \tag{79}$$

$$\epsilon_{r\theta} = 0 \tag{80}$$

To account for large strains, first order nonlinear terms are retained.

6.3 Strain Energy

The strain energy expression can be written as

$$U = 1/2 \int \sigma\epsilon dV \tag{81}$$

and for the stress-strain components considered, the expression for the strain energy is

$$U = \frac{1}{2} \int_0^{2\pi} \int_0^a \int_{-h/2}^{-h/2} (E_{rr}\epsilon_r^2 + E_{\theta\theta}\epsilon_\theta^2 + 2E_{r\theta}\epsilon_r\epsilon_\theta) r \, dz \, dr \, d\theta \quad (82)$$

The strain energy of the plate configuration, as shown schematically, is taken as the sum of the elemental strain energies of the plate. Thus,

$$U = \frac{1}{2} \sum_{k=1}^{N_\theta} \sum_{j=1}^{N_R} \int_{-h/2}^{h/2} (E_{rr}\epsilon_r^2 + E_{\theta\theta}\epsilon_\theta^2 + 2E_{r\theta}\epsilon_r\epsilon_\theta) dz \left(\frac{r_{j+1}^2 - r_j^2}{2} \right) (\theta_{k+1} - \theta_k)$$

$$(83)$$

It should be noted that for the case when N_θ and N_R are large, the summation equals the integral equation.

The impact load distribution is taken to be elliptical and of the form

$$q(r) = \frac{3P}{2\pi a_c^2} \left[1 - \left(\frac{r}{a_c} \right)^2 \right]^{1/2} \quad (84)$$

where P = the impact load and a_c = the radius of contact between the sphere and the plate.

The potential energy for the plate loading $q(r)$ is taken as

$$V = - \int_0^{a_c} \int_0^{2\pi} q(r) w r \, dr \, d\theta \quad (85)$$

Using the expressions for $q(r)$ and $w(r)$, V becomes

$$V = -Pw_o \left[1 - .67515 \left(\frac{a_c}{a} \right) + .8 \left(\frac{a_c}{a} \right)^2 \ln \left(\frac{a_c}{a} \right) \right] \quad (86)$$

For a_c small, $V = -Pw_o$.

The total potential energy π of the deformed plate is, $\pi = U + V$.

Taking the expressions for U and V, the three unknown constants $C_1, C_2,$

and w_o can be found by minimizing the total potential energy expression, thus,

$$A_{11}C_1 + A_{12}C_2 = -B_1 - D_1 \tag{87}$$

$$A_{12}C_1 + A_{22}C_2 = -B_2 - D_2 \tag{88}$$

$$A_{33}(C_1,C_2,w_o) = P\left[1 - .67515\left(\frac{a_c}{a}\right)^2 + .8\left(\frac{a_c}{a}\right)^2 \ln\left(\frac{a_c}{a}\right)^2\right] \tag{89}$$

The constants are defined as follows:

$$A_{11}\sum_{i=1}^{N_L}\sum_{j=1}^{N_R}\sum_{k=1}^{N_\theta}(z_{i+1} - z_i)A_{11}'\Delta_{jk} \tag{90}$$

$$A_{12} = \sum\sum\sum(z_{i+1} - z_i)A_{12}'\Delta_{jk} \tag{91}$$

$$A_{22} = \sum\sum\sum(z_{i+1} - z_i)A_{22}'\Delta_{jk} \tag{92}$$

$$B_1 = \sum\sum\sum\left(\frac{z_{i+1} - z_i}{2}\right)B_1'\Delta_{jk} \tag{93}$$

$$B_2 = \sum\sum\sum\left(\frac{z_{i+1} - z_i}{2}\right)B_2'\Delta_{jk} \tag{94}$$

$$D_1 = \sum\sum\sum\left(\frac{z_{i+1} - z_i}{2}\right)D_1'\Delta_{jk} \tag{95}$$

$$D_2 = \sum\sum\sum\left(\frac{z_{i+1} - z_i}{2}\right)D_2'\Delta_{jk} \tag{96}$$

$$A_{33} = \sum\sum\sum A_{33}'\Delta_{jk} \tag{97}$$

where

$$A_{11} = (a - 2\bar{r})^2 E_{\bar{r}\bar{r}} + (a - \bar{r})^2 E_{\theta\bar{\theta}} + 2E_{\bar{r}_\theta}(a - \bar{r})(a - 2\bar{r})$$

$$- \left(\frac{z_{i+1} + z_i}{2}\right)(a - 2_{\bar{r}})^2\left(E_{\bar{r}\bar{r}}w'' + \frac{E_{\bar{r}\theta}}{\bar{r}}w'\right)$$

$$A'_{22} = \bar{r}^2(2a - 3\bar{r})^2 E_{\bar{r}\bar{r}} + \bar{r}^2(a - \bar{r})(2a - 3\bar{r})E_{\bar{r}\theta}$$

$$+ \bar{r}^2(a - \bar{r})^2 E_{\theta\theta} - \left(\frac{z_{i+1} + z_i}{2}\right)\bar{r}(2a - 3\bar{r})^2(E_{\bar{r}\bar{r}}w'' + E_{\bar{r}\theta}w')$$

$$A'_{12} = \bar{r}(a - 2\bar{r})(2a - 3\bar{r})E_{\bar{r}\bar{r}} + \bar{r}(a - \bar{r})(3a - 5\bar{r})E_{\bar{r}\theta}$$

$$+ \bar{r}(a - \bar{r})^2 E_{\theta\theta} - \left(\frac{z_{i+1} + z_i}{2}\right)(a - 2\bar{r})(2a - 3\bar{r})$$

$$\times (E_{\bar{r}\bar{r}}\bar{r}w'' + E_{\bar{r}\theta}w')$$

$$A'_{33} = \left(\frac{z_{i+1} - z_i}{2w_o}\right)$$

$$\times \left\{ E_{\bar{r}\bar{r}}[4(u')^2 + 2(u')^4 + 6(u')^3 + 3(u'w')^2 + 4u'(w')^3 + (w')^4]\right.$$

$$+ 2E_{\bar{r}\theta}\left[4\left(\frac{u}{\bar{r}}\right)u' + 2\left(\frac{u}{\bar{r}}\right)(w')^2 + 3\left(\frac{u}{\bar{r}}\right)(u')^2\right]4] - \left(\frac{z_{i+1}^2 - z_i^2}{2}\right)$$

$$\times \left(\frac{E_{\bar{r}\bar{r}}}{w_o}\left[3u' + \frac{5}{2}(u')^2 + \frac{3}{2}(w')^2\right]w'' + \frac{3u}{\bar{r}^2 w_o}(w')E_{\theta\theta}\right)$$

$$+ E_{\bar{r}\theta}\left\{\frac{3u}{w_o\bar{r}}w'' + (2a - 3\bar{r})w'[1 + (a_1 - 3\bar{r})c_1 + \bar{r}(2a - 3\bar{r})c_2]\right\}$$

$$+ \frac{16w_o}{a^4}\left(\frac{z_{i+1} + z_i}{3}\right)\left[E_{\bar{r}\bar{r}}\left(1 + \ln \frac{\bar{r}}{a}\right)^2 + E_{\theta\theta}\ln \frac{\bar{r}}{a}\left(1 + \ln \frac{\bar{r}}{a}\right)\right]$$

$$B'_1 = [(a - 2\bar{r})E_{\bar{r}\bar{r}} + (a - \bar{r})E_{\bar{r}\theta}](w')^2 + (z_{i+1} + z_i)$$

$$\times \left\{E_{\bar{r}\bar{r}}(a - 2\bar{r})w'' + E_{\theta\theta}\left(\frac{a}{\bar{r}}\right)w'E_{\bar{r}\theta}\left[(a - \bar{r})w'' + \left(\frac{a - 2\bar{r}}{\bar{r}}\right)w'\right]\right\}$$

$$B_2' = [\bar{r}(2a - 3\bar{r})E_{\bar{r}\bar{r}} + \bar{r}(a - \bar{r})E_{\bar{r}\theta}](w')^2 + (z_{i+1} + z_i)$$

$$\times \{E_{\bar{r}\bar{r}}(2a - 3\bar{r})\bar{r}w'' + E_{\theta\theta}(a - \bar{r})w'$$

$$+ E_{\bar{r}\theta}[(a - \bar{r})w'' + (2a - 3\bar{r})w']\}$$

$$D_1' = E_{\bar{r}\bar{r}}(a - 2\bar{r})u'[(u')^2 + 3u' + (w')^2]$$

$$+ E_{\bar{r}\theta}u'\left[(a - \bar{r})u' + \frac{2u}{\bar{r}}(a - 2\bar{r})\right]$$

$$D_2' = E_{\bar{r}\bar{r}}(2a - 3\bar{r})\bar{r}u'[(u')^2 + 3u' + (w')^2]$$

$$+ E_{\bar{r}\theta}u'[\bar{r}(a - \bar{r}) + 2u(2a - 3\bar{r})]$$

and

$$\Delta_{jk} = \frac{r_{j+1}^2 - r_j^2}{2}(\theta_{k+1} - \theta_k)$$

$$\bar{r} = \bar{r}_j = \left(\frac{r_{j+1} + r_j}{2}\right), \qquad \bar{\theta} = \bar{\theta}_k = \left(\frac{\theta_{k+1} + \theta_k}{2}\right)$$

$$u' = C_1\frac{w_o^2}{a^3}(a - 2\bar{r}) + C_2\frac{w_o^2}{a^4}(2a\bar{r} - 3\bar{r}^2)$$

$$w' = \frac{4w_o}{a}\left(\frac{\bar{r}}{a}\right)\ln\left(\frac{\bar{r}}{a}\right), \qquad w = \frac{4w_o}{a^2}\left(\ln\frac{\bar{r}}{a} + 1\right)$$

Two types of damage have been evaluated using this approach:

- splitting damage – due to transverse tensile stresses or in plane shear stresses
- fiber failure – due to longitudinal stresses in the lamina

A numerical procedure has been used to evaluate the three equations involving C_1, C_2, P, and a_c. The procedure used is as follows:

- Select an initial value of w_o.
- Select increments for w_o.
- Calculate C_1 and C_2.
- Calculate P.
- Check for failure using the Tsai-Wu criterion and/or Maximum-Stress criterion.
- If no failure, repeat calculations.

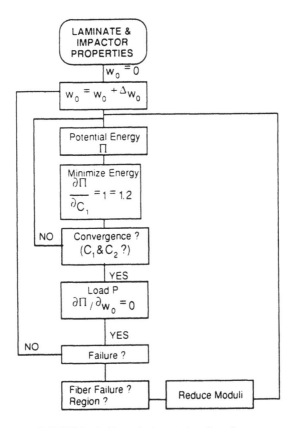

FIGURE 2.15. Numerical procedure flow diagram.

- Calculate region of splitting failure.
- Calculate width of the fiber failure region.
- Reduce E_l, G_{lt}, to zero in the splitting damage region.
- Reduce E_l, E_t, and G_{lt} to zero in the region of both splitting and fiber damage.
- Repeat the calculation procedure.

A flow diagram of the numerical procedure used is shown in Figure 2.15.

6.4 Summary of Model V

Results have been obtained and presented in terms of load versus deflection curves and damage plots. Predicted load deflection data for three plate

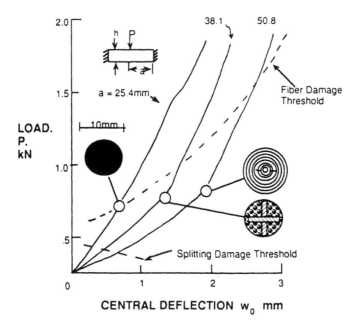

FIGURE 2.16. Diagram of damage thresholds.

radii are shown below. In these curves, the splitting and fiber damage thresholds have been indicated. Some important observations include (see Figure 2.16).

- Larger plates have a higher *fiber damage threshold* than small plates.
- Larger plates have a lower *splitting threshold* then small plates.
- Larger plates (for a given load and thickness) develop larger damage regions.

Additional conclusions are as follows:

- The first failure mode was splitting and is initiated in the bottom-most ply.
- The splitting damage region in each ply is elongated along the fiber direction.
- First fiber failure thresholds (for a given load or energy) are higher for large plates than for small plates over a limited range of radius to thickness ratios ($a/h > 10$).

MODEL VI

Reference:	Husman, Whitney, Halpin [12]
Model classification:	residual strength degradation
Impact velocity regime:	low, intermediate, high
Striker/target characteristics:	rigid/flexible

EXPERIMENTAL PARAMETERS

Striker	Target
mass – small	mass – large
material – steel	material – B/Ep, Gl/Ep, Gr/Ep
geometry – spherical	geometry – rectangular plate
striker incidence – normal	thickness – thin, intermediate
velocity – low to high	ply lay-up – orthotropic
	boundary conditions – clamped

7. ANALYSIS—MODEL VI

In this model a procedure for converting the impact damage to an equivalent crack of known dimensions is developed for use in evaluating the residual strength.

Consider a slit of length $2c$ embedded in an orthotropic plate carrying a uniform tensile stress $\bar{\sigma}$ applied at infinity and perpendicular to the slit. For such a plate, it has been shown that the critical strain energy release rate G_{Ic} takes the form

$$G_{Ic} = K_{Ic}^2 \left\{ \left(\frac{\bar{S}_{11}\bar{S}_{22}}{2} \right) \left[\left(\frac{\bar{S}_{22}}{\bar{S}_{11}} \right)^{1/2} + \frac{2\bar{S}_{12} + \bar{S}_{66}}{2\bar{S}_{11}} \right] \right\}^{1/2} \tag{98}$$

where K_{Ic} are the critical stress intensity factor and $\bar{S}_{ij}$ are the orthotropic plate compliances.

The value of K_{Ic} for an orthotropic material is the same as that for an isotropic material, that is,

$$K_{Ic} = \bar{\sigma}\sqrt{\pi c} \tag{99}$$

The equation for G_{Ic} can be rewritten using the result for K_{Ic} as,

$$G_{Ic} = A c \bar{\sigma}^2 \frac{\bar{S}_{22}}{2} \tag{100}$$

with

$$A = \pi \left\{ \frac{2\bar{S}_{11}}{\bar{S}_{22}} \left[\left(\frac{\bar{S}_{22}}{\bar{S}_{11}} \right)^{1/2} + \frac{2\bar{S}_{12} + \bar{S}_{66}}{2\bar{S}_{11}} \right] \right\}^{1/2}$$

For the case of an isotropic material, the value of A reduces to 2π.

In the expression for G_{Ic}, the quantity $(\bar{\sigma}^2 \bar{S}_{22})/2$ represents the work/unit volume (W_b), and is the area under the stress-strain curve at a distance from the slit. Thus, G_{Ic} can be written as

$$G_{Ic} = A_c W_b \tag{101}$$

For an isotropic material the above expression is given by

$$G_{Ic} = 2\pi c W_b \tag{102}$$

It is assumed that a damage zone exists in proximity to the inherent stress concentration, and that this damage zone comprises a characteristic volume of material, which must be stressed to a critical level before fracture occurs. The damage zone is assumed to be identified by a characteristic dimension c_o, which can be considered as an effective flaw length. This dimension is considered to be a control for the strength of the composite. Thus, the critical strain energy release rate expression can be written as

$$G_{Ic} = A c_o W_s \tag{103}$$

The quantity W_s, as introduced, is the strain energy under the stress-strain curve for the case of a statically loaded composite without a mechanically implanted flaw. Considering an effective half crack length $c + c_o$, the strain energy release rate for an orthotropic plate can then be written as

$$G_{Ic} = A(c + c_o) W_b \tag{104}$$

An expression relating the flawed and unflawed strength can thus be written as

$$\sigma_R = \sigma_o \sqrt{\frac{c_o}{c + c_o}} \qquad (105)$$

An analogy can now be developed for the case of hard particle impact on a composite target to that of damage imparted by implanting a crack of fixed dimensions in a tensile coupon. It is first assumed that the difference required to break an undamaged specimen, as opposed to the energy required to break an impacted specimen, is proportional to the kinetic energy imparted to the specimen over a representative volume of the specimen. Thus,

$$W_s - W_b = k \frac{W_{KE}}{V} \qquad (106)$$

and V is the representative volume. This volume can be characterized by a surface area A_e and plate thickness t. The area A_e is assumed to be independent of the kinetic energy. Thus, an energy balance expression can be written as

$$W_s - W_b = K \overline{W}_{KE} \qquad (107)$$

where $K = k/A_e$ and $\overline{W}_{KE} = W_{KE}/t$.

A relationship can thus be established between impact damage and that associated with a mechanically induced crack, that is,

$$G_{Ic} = A(c + c_o)(W_s - K \overline{W}_{KE}) \qquad (108)$$

Recalling that

$$G_{IC} = Ac_o W_s \qquad (109)$$

the two expressions can be equated, and solving for c yields

$$c = \frac{c_o K \overline{W}_{KE}}{(W_s - K \overline{W}_{KE})} \qquad (110)$$

The corresponding residual strength can then be expressed in terms of the kinetic energy as

$$\sigma_R = \sigma_o \sqrt{\frac{W_s - K\overline{W}_{KE}}{W_s}} \tag{111}$$

This expression implies that the residual strength can be obtained through two experiments:

- a static tensile test on an unflawed specimen
- a static tensile test on an impacted specimen

The factor K is related to specimen geometry, laminate stacking sequence, and boundary conditions.

A key parameter in the above equation is W_s, which can be determined by theoretical methods. This is accomplished through calculating the area under the stress-strain diagram for the specimen. A flow chart of the calculation procedure is shown (see Figure 2.17).

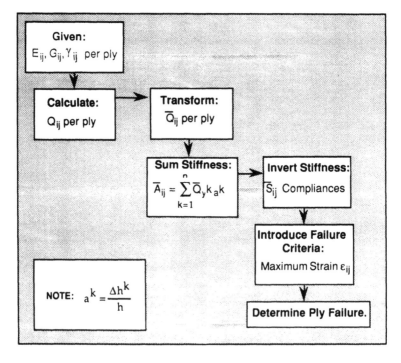

FIGURE 2.17.

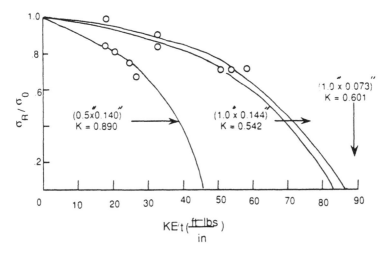

FIGURE 2.18. Effect of specimen size on impact residual strength.

Figure 2.17 can be used to determine W and with knowledge of the striker and target; that is, striker mass, velocity, and target thickness the residual target strength can be evaluated.

For velocities greater than the velocity required to perforate the target, the residual strength is reduced to that of a target with a hole and, thus, is independent of the impact event. Therefore, a lower threshold limit exists for the retained residual strength for such an event. For velocities less than this limit, the residual strength is monotonically reduced. This is shown schematically in Figure 2.18.

Additional strength reduction data has been obtained and presented for several parameters, including

- effect of matrix
- effect of fiber
- effect of specimen size
- effect of specimen width to striker diameter

As an example, the effect of kinetic energy delivered to the target and specimen size on the residual strength of the target are shown in Figures 2.18 and 2.19.

7.1 Summary of Model VI

(1) The type of matrix used in the composite configuration (thermoset ver-

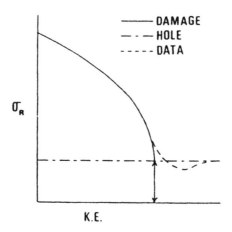

FIGURE 2.19. Residual strength as a function of kinetic energy.

sus thermoplastic) has little effect on the residual strength after impact; however, it does effect the damage mode. Some of these modes are identified as

- thermoset (local crushing, matrix cracking, delamination)
- thermoplastic (indentation)

(2) The effect of fiber type as used in the composite configuration is important for the retained residual strength of the composite after impact. High strength, high strain fibers are found to provide higher resistance to impact damage.

(3) The effect of specimen size is also important in assessing the residual strength of an impact damaged composite. It appears that in the design of experiments to evaluate the residual strength of impact damaged composites, the thickness effect may be removed as a control parameter for specimen width to striker diameters greater than five.

MODEL VII

Reference:	El-Zein, Reifsnider [13]
Model classification:	residual strength degradation
Impact velocity regime:	low
Striker/target characteristics:	rigid/flexible

EXPERIMENTAL PARAMETERS

Striker	Target
mass – large	mass – small
material – steel	material – Gr/Ep
geometry – hemispherical	geometry – square panels
striker incidence – normal	thickness – thin
velocity – low	ply lay-up – orthotropic, quasi- isotropic
	boundary conditions – simply supported

---

8. ANALYSIS—MODEL VII

The Husman et al. model [12] discussed previously has focused attention on the evaluation of the retained compressive strength of an impacted composite laminate as a function of the kinetic energy of the impact event.

The model discussed in this section focuses on the prediction of the tensile residual strength after impact, as a function of the damaged area created by the impact event. For analysis, the damaged area is simulated as an elliptical inclusion with a solution generated based upon a complex variable formulation. A description of this methodology follows:

For an anisotropic plate with an elliptical inclusion subjected to tensile stresses, the global stresses can be defined in terms of two complex functions, $\phi_1(z_1)$, $\phi_2(z_2)$,

$$\sigma_x = p + 2R_e[\mu_1^2 \, \phi_1'(z_1) + \mu_2^2 \, \phi_2'(z_2)] \qquad (112)$$

$$\sigma_y = q + 2R_e[\phi_1'(z_1) + \phi_2'(z_2)] \qquad (113)$$

$$\tau_{xy} = t - 2R_e[\mu_1\phi_1'(z_1) + \mu\phi_2'(z_2)] \qquad (114)$$

The quantities p, q, and t appearing on the right-hand side of the stress components represent the applied stresses while μ_1 and μ_2 are roots of the following equation:

$$a_{11}\mu^4 - 2a_{16}\mu^3 + (2a_{12} + a_{66})\mu^2 - 2a_{26}\mu + a_{22} = 0 \qquad (115)$$

The quantities a_{ij} are elements of the compliance matrix of the plate while the complex functions $\phi_1(z_1)$ and $\phi_2(z_2)$ are given by

$$\phi_1(z_1) = \frac{1}{2(\mu_1 - \mu_2)}[(A - p)bi - (B - q)\mu_*a$$
$$+ (C - t)(i\mu_2 b - a)]\frac{1}{\zeta_*} \tag{116}$$

$$\phi_2(z_2) = \frac{1}{2(\mu_1 - \mu_2)}[(A - p)bi - (B - q)\mu_1 a$$
$$+ (C - t)(i\mu_1 b - a)]\frac{1}{\zeta_2} \tag{117}$$

The quantities ζ_1, ζ_2 are

$$\zeta_i = \frac{Z_1 + \sqrt{Z_i^2 - a^2 - \mu_i^2 b^2}}{a - i\mu_i b} \qquad i = 1,2 \tag{118}$$

Also, a and b are the radii of the elliptical inclusion, while A, B, and C are tensile stresses and shear stresses in the inclusion. These quantities are determined using continuity conditions between the plate and the inclusion.

The elements of the compliance matrix, that is, the a_{ij}, are associated with the inclusion and the plate. This relationship can be written as

$$a_{ij}' = ma_{ij} \tag{119}$$

Here the, a_{ij}' are related to the inclusion factor and the a_{ij} are related to the plate. The factor m is zero for a rigid inclusion and infinite for the case of an open hole.

Once the global stresses, σ_G, have been determined, the in-plane forces can be found using

$$\{N\} = \{\sigma_G\}H \tag{120}$$

where H represents the thickness of the laminate.

The laminate strains, in turn, can be found once the in-plane loads have been determined. Thus,

$$\{\epsilon\} = [A]^{-1}\{N\} \tag{121}$$

where $[A]$ is the extensional stiffness matrix.

Having determined the laminate strains, the laminate stresses can be found from:

$$\{\sigma_p\} = [\bar{Q}]\{\epsilon\} \tag{122}$$

where $[\bar{Q}]$ is the reduced stiffness matrix and $\{\sigma_p\}$ are the ply stresses.

The residual strength of the composite laminate can then be determined using the result

$$\frac{1}{D_o} \int_b^{b+D_o} \frac{\sigma_{ip}(0,y)}{X_T} dy = 1 \tag{123}$$

In the above equation, σ_{ip} are the ply stresses, X_T is the unidirectional tensile strength, and D_o is a characteristic dimension over which the stresses are integrated.

An experimental program was developed to evaluate the residual strength of impacted composite laminates in support of the theoretical model. The following equipment and specimens were used:

(1) Impact System – Dynatup Model 820
(2) Data Recording System – General Research Corporation 730-I data system
(3) Target Construction – Laminate
(4) Target Fabrication – Autoclave
(5) Target Ply Configuration – $[(0/90)_4]_s$, $[0/\pm45/90]_{2s}$, $[0/\pm45/0]_{2s}$
(6) Target Geometry – 6″ × 6″
(7) Target Thickness – 8 plies
(8) Target Material – Graphite/Epoxy

To determine the residual strength of the impacted laminate, it is necessary to obtain information on the factor m, that is, the ratio of the compliance of the plate to that of the inclusion. To obtain this relationship it can be argued that the ratio of the plate material to the inclusion material can be related as the ratio of a damaged area with respect to a reference area. Thus,

$$m = \frac{a'_{ij}}{a_{ij}} = \frac{A_D}{A_{ref}}$$

The reference area is likened to the concept that, up to a characteristic flaw size, there is no significant strength change. The damage area is thus taken as that obtained at a specific energy level.

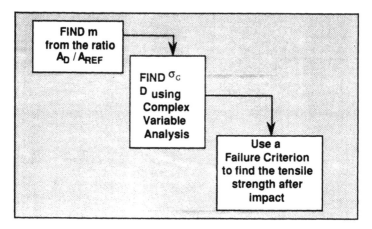

FIGURE 2.20.

Once the value of m has been determined, the predictive procedure to find the residual strength can be obtained using the flow chart in Figure 2.20.

8.1 Summary of Model VII

An approach to the prediction of the tensile residual strength of an impacted composite laminate has been developed. The technique is based upon experimentally determined matrices of the impact damaged composite and the subsequent reduction of the composite stiffness in this region. This, in turn, is used as input to determine stresses in the composite, which are then used to examine ply level failure. The approach used provides insight into the effect of the impact event, that is, that such effects are the result of an abrupt change in the elastic properties of the damaged region and are not related to the loss in inherent strength associated with the damage. Some factors requiring further work include consideration of through-the-thickness variations in damage, thickness of the composite, stacking sequence effects, as well as the residual strength for testing modes other than tension.

MODEL VIII

Reference: Lal [14]
Model classification: deformation mechanics
Impact velocity regime: low
Striker/target characteristics: hard/soft

EXPERIMENTAL PARAMETERS	
Striker	Target
mass – large	mass – small
material – steel	material – Gr/Ep
geometry – spherical	geometry – circular plate
striker incidence – normal	thickness – thin
velocity – low	ply lay-up – quasi-isotropic
	boundary conditions – clamped

9. ANALYSIS—MODEL VIII

An energy balance model has been introduced to determine the energy absorbed by delaminations due to impact. The methodology involved in computing the delamination energy requires information on the load-deflection characteristic of the target during the impact event. The phenomena involved during loading are

- flexural and membrane effects
- Hertzian contact effect
- shear deformation effects

These effects can be expressed analytically in terms of the structural and material stiffnesses associated with these phenomena and are mechanically simulated by three springs. Considering the simultaneous action of each of the springs, the stiffnesses can be simulated by three springs in a series (Figure 2.21). This leads to the following expression:

$$K_e = \left[\frac{1}{k_i} + \frac{1}{k_f} + \frac{1}{k_s} \right]^{-1} \tag{124}$$

where K_e is the equivalent stiffness; k_i is the indentation stiffness, k_f is the flexural stiffness, and k_s is the shear stiffness.

The energy and force relations for the case of large deflections of plates are given, respectively, by

$$I = \frac{2Eh^5}{\alpha D^2} \left[\left(\frac{\delta}{h} \right)^2 + \frac{1}{2}\beta \left(\frac{\delta}{h} \right)^4 \right] \tag{125}$$

$$P = \frac{4Eh^4}{\alpha D^2} \left[\left(\frac{\delta}{h} \right) + \beta \left(\frac{\delta}{h} \right)^3 \right] \tag{126}$$

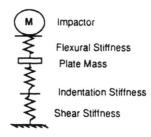

Impactor

Flexural Stiffness

Plate Mass

Indentation Stiffness

Shear Stiffness

FIGURE 2.21. Equivalent stiffness model during impact.

where

E = the plate elastic modulus
h = the plate thickness
d = the plate diameter
α = a geometry constant
β = a membrane parameter

For example, for thin plates of the order of eight plies, the quantity β is expressed in terms of the coefficient restitution e as

$$\beta = .443e^2 \tag{127}$$

Assuming a shear dominated theory for the composite laminate and that the shear behavior is elastic-plastic during loading and elastic during unloading, the coefficient of restitution can be plotted as a function of the impact velocity. The impact force can be approximated from the first mode of the vibration event as

$$P = V(MK)^{1/2} \tag{128}$$

A curve of the impact force versus deformation can be obtained for a specified impact velocity, neglecting plate vibrations equivalent to a static loading condition. This is graphically displayed in Figure 2.22.

- Curve A represents elastic loading.
- Curve B represents the loading during progressive plate delamination.
- Curve C represents elastic unloading.

The area bounded by curves B and C represents the energy absorbed by delamination.

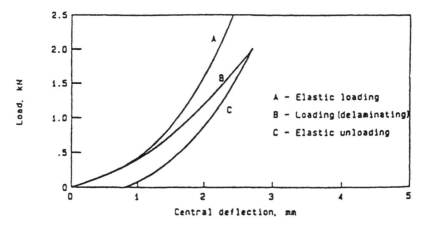

FIGURE 2.22. Energy absorbed schematic.

The fiber breakage energy can be expressed by the following equation:

$$I_f = I_a - I_d \qquad (129)$$

where I_a is the net energy absorbed by the plate and I_d is the delamination.

The net energy absorbed can be obtained by use of the coefficient of restitution as

$$I_a = I(1 - e^2) \qquad (130)$$

The respective energy distributions associated with each of these effects is graphically shown in Figure 2.23.

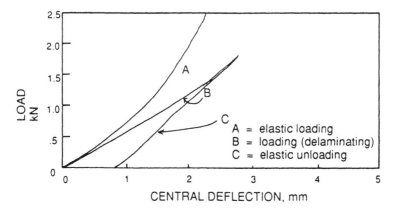

FIGURE 2.23. Energy distribution diagram.

It should be noted that energy losses due to indentation and air drag are neglected. The above discussion focuses on an energy balance between the principal mechanisms of energy dissipation for an impacted composite specimen. In addition to the energy balance model, the ability of the composite specimen to resist impact damage and to retain its undamaged strength are at issue. The basis for an approach to solving this problem is to make use of linear elastic fracture mechanics (LEFM). Starting with the classical fracture criteria for an orthotropic material, the associated strain energy release rate necessary to create a unit fracture surface area can be written as

$$G_{1c} = K_{lc}^2 \left[\left(\frac{S_{11} E_{22}}{2} \right) \left\{ \left(\frac{S_{22}}{S_{11}} \right)^{1/2} + \frac{2S_{12} + S_{66}}{2S_{11}} \right\} \right] \tag{131}$$

where K is the critical stress intensity factor and the S_{ij} are the orthotropic plate compliances.

For a quasi-isotropic material, the above expression can be written as

$$G_{lc} = K_{lc}^2 / E \tag{132}$$

with E representing the laminate modulus. Considering a slit in the orthotropic material, the value of K_{lc} can be expressed as

$$K_{lc} = \sigma_r \sqrt{\pi(L + L_o)} \tag{133}$$

with σ_r the residual strength of the plate having a center slit $2L$, and L_o the inherent flaw size. For the case of an unflawed specimen, the value of K_{lc} is expressed as

$$K_{lc} = \sigma_o \sqrt{\pi L_o} \tag{134}$$

with σ_o the tensile strength of the unflawed specimen. Equating the two respective stress intensity functions gives

$$\sigma_r = \sigma_o \sqrt{L_o / (L + L_o)} \tag{135}$$

The equivalent slit length for an impact damaged specimen can be obtained from knowledge of the strain energy release rate and the fiber fracture energy given by

$$L = I_f/(2hG_{Ic})$$ (136)

The residual strength of the impact damaged specimen can be found using the above expression and fracture toughness data for σ_o, L_o substituted into the residual strength expression. An examination of this expression suggests that, for the case of brittle materials, in order to increase the strain energy release rate or work/unit volume for impact damage resistance, a higher strain to failure composite is needed. The general trend of residual strength reduction with impact energy is shown in Figure 2.24.

To test the model, damage was introduced by means of a dropped steel ball on the test specimen. The residual strength of the damaged specimen was determined in a tensile mode based upon an evaluation of the damage state using ultrasonic scanning. In addition, the inherent flaw size and the specimen fracture toughness were found. Using this data, the net energy absorbed by the plate was determined, as was the fiber fracture energy. The experimental data and analytical predictions are found to be in agreement

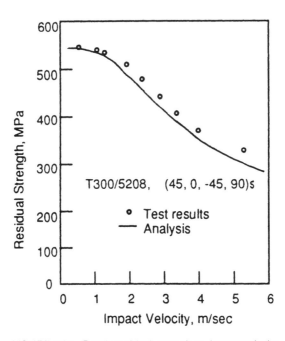

FIGURE 2.24. Tensile residual strength vs. impact velocity.

TABLE 2.4.

Reference	Problem Area	Striker Type	Target Type	Model Approach	Predictor
Greszczuk	Deformation mechanics	Spherical	Semi-infinite media	Hertzian contact	Impact duration, force time history
Shivakumar, Elber, Illg	Deformation mechanics	Spherical	Circular plate	Energy balance, spring-mass	Impact force magnitude, force time history
Cairns, Lagace	Deformation mechanics	Spherical	Square plate	Hertzian contact, energy balance	Impact force
Madsen, Morgan, Nuismer	Deformation mechanics	Spherical	Spherical lumped mass	Spring-mass	Impact time duration
Shivikumar, Elber, Illg	Damage mechanics	Spherical	Circular plate	Strain energy	Splitting damage fiber failure
Husman, Whitney, Halpin	Residual strength degradation	Spherical	Rectangular plate	Fracture mechanics	Residual tensile strength
El-Zein, Reifsnider	Residual strength degradation	Hemispherical	Square plate	Energy balance	Residual tensile strength
Lal	Residual strength degradation	Spherical	Circular plate	Energy balance	Residual tensile strength

within 15%. A comparison of the tensile residual strength data obtained with residual strength data reported in the literature for the case of rectangular plates is shown to follow a similar trend.

9.1 Summary of Model VIII

For the impact velocity regimes considered in the experiments conducted and for the model developed, the impact damage consisted primarily of delamination and fiber breakage. Considering the case of large deflections of impacted specimens, a membrane parameter related to the coefficient of restitution was found to be important for characterizing the event. The load deflection response of the impacted composite specimen has been described in terms of a spring mass model, and the resultant change in stiffness of the system described by laminate debonding. This, in turn, leads to a methodology for evaluating the energy absorbed due to this important mechanism. Further, by examining the loading and unloading cycle of the event, the net energy absorbed can be determined. This can then be related to the fiber breakage energy. The experimental and analytical data suggested that, for developing impact resistant composite materials, a higher strain to failure for the composite system is desirable.

10. CONCLUDING REMARKS

The analytical models described above are representative of the current state of investigation for evaluating the impact dynamics of composite materials. While not described in this chapter, the contribution of A. L. Dobyns [15] is recognized as providing an early incentive to the development of low velocity impact expressions. As a means of summarizing the contributions of all the authors cited in the models described, Table 2.4 has been prepared. This table, as well as the preceding discussion, should be useful to those investigators charged with studying this subject.

11. REFERENCES

1. Demuts, E. 1989. "Damage Tolerance of Composites," *Proceedings of the American Society for Composites*, 425–433.
2. McQuillen, E. J., L. W. Gause and R. E. Llorens. 1976. "Low Velocity Transverse Normal Impact of Graphite Epoxy Composite Laminates," *Journal Composite Materials*, 10:79–91.
3. Riera, J. D. 1982. "Basic Concepts and Load Characteristics in Impact Problems," in *Concrete Structures under Impact and Impulsive Loading*, Berlin, pp. 9–12.

4. Milton, J. E., R. L. Sierakowski, C. W. Bobbitt and C. C. Schauble. 1980. AFOSR Rept. 79-0071, pp. 29–30.

5. Greszczuk, L. B. 1975. "Response of Isotropic and Composite Materials to Particle Impact," *ASTM STP 568*, pp. 183–211.

6. Greszczuk, L. B. and H. Chao. 1977. "Impact Damage in Graphite Fiber Reinforced Composites," *ASTM STP 617*, pp. 389–408.

7. Greszczuk, L. B. 1982. "Damage in Composite Materials Due to Low Velocity Impact," *Impact Dynamics*, A. A. Zukas et al. eds., New York: J. Wiley, pp. 55–94.

8. Shivakumar K. N., W. Elber and W. Illg. 1985a. "Prediction of Impact Force and Duration Due to Low-Velocity Impact on Circular Composite Laminates," *J. Appl. Mech.*, pp. 52:674–680.

9. Cairns, D. S. and P. A. Lagace. 1989. "Transient Response of Graphite/Epoxy and Kevlar/Epoxy Laminates Subjected to Impact," *AIAA Journal*, 27(11):1590–1596.

10. Madsen, C. B., M. E. Morgan and R. J. Nuismer. 1990. "Scaling Impact Response and Damage in Composites. Damage Assessment for Composite Cases," Air Force Technical Report, AL-TR-90-037, Vol. 2.

11. Shivakumar K., W. Elber and W. Illg. 1985b. "Prediction of Low-Velocity Impact Damage in Thin Circular Laminates," *AIAA Journal*, 23(3), pp. 442–449

12. Husman, G. E., J. M. Whitney and J. C. Halpin. 1975. "Residual Strength Characterization of Laminated Composites Analyted to Impact Loading," *ASTM STP 568*, pp. 92–113.

13. El Zein, M. S. and K. G. Reifsnider. 1990. "On the Prediction of Tensile Strength after Impact of Composite Laminates," *Journal of Composites Technology and Research*, 12, pp. 147–154.

14. Lal, K. M. 1983. "Residual Strength Assessment of Low Velocity Impact Damage of Graphite-Epoxy Laminates," *Journal of Reinforced Plastics and Composites*, 2, pp. 226–238.

15. Dobyns, A. L. 1981. "Analysis of Simply-Supported Orthotropic Plates Subject to Static and Dynamic Tools," *AIAA Journal*, 19:642–650.

CHAPTER 3

Damage Tolerance Evaluation

1. INTRODUCTION

A good understanding of damage is essential to develop a rational methodology to address damage tolerance characteristics of advanced composite materials. Because composite materials are made of different types of constituent materials and often have different lay-ups for varied aircraft structural applications, a multitude of defects are found in these material systems, which are primarily attributable to fabrication/processing (birth defects) and in-field/service problems. The typical defects and sources of damage are listed as follows.

1.1 Typical Defects in Composites

1.1.1 TYPES OF DEFECTS

- disbonds
- delaminations
- inclusions (bugs, foreign contamination, etc.)
- voids, blisters
- impact damage (tool drop, pebble)
- fiber misalignment
- cut or broken fibers (dents)
- abrasions, scratches
- wrinkles
- resin cracks, crazing
- density variations
- improper cure (soft resin)
- matching problems (improper hole size, etc.)

113

1.2 Sources of Damage

1.2.1 FABRICATION/PROCESSING (BIRTH DEFECTS)

Fabrication/processing damage (defects) generally include

- abrasions, scratches, dents, punctures
- cut fibers
- knots, kinks
- improper splicing
- voids (due to poor processing, highly advanced resin, excess humidity/moisture)
- resin-rich, resin-lean areas (improper tensioning)
- subquality materials
- cure problems (uncured resin)
- inclusions, bugs, foreign contamination, etc.
- tool installation and removal during processing
- mandrel removal problems (handling)
- machining problems
- shipping to propellant processing
- tool drop (impact damage)
- proof testing (crazing)

1.2.2 IN-FIELD/SERVICE PROBLEMS

In-field/service problems generally include

- vibration
- shock
- lightning damage
- environmental cycling (temperature and humidity)
- flight loads (fatigue)
- improper repair (maintenance)
- in-storage creep or handling loads
- pebble impact (tool drop)
- scratches, dents, punctures
- corrosion
- erosion, dust, sand
- bacterial degradation

Damage tolerance evaluation is essential to determine the load carrying capability of the material for a specific structural application. The performance capability of the composite in the presence of flaws is a measure of structural integrity. Therefore, data obtained from a damage tolerance assessment is not only valuable in selecting material systems, it is also useful to assess the structural integrity of components. The aircraft industry has developed, over the years, a number of damage evaluation techniques and mechanical tests to assess the integrity of structural composites. These procedures are addressed in a chronological fashion in this chapter.

This chapter is organized into three sections, Nondestructive Evaluation Techniques, Damage Assessment Using Optical/SEM Evaluation, and Damage Tolerance Tests. The first two sections are adapted from the interim 1987 Boeing Report #AFWAL-TR-86-4137 titled "Compendium of Post-Failure Analysis Techniques for Composite Materials" [1]. The last section is adapted from NASA 1092 standard (1982) [2]. These areas provide a comprehensive view of the type of defects and damage commonly encountered in advanced structural composites, techniques to evaluate them, and mechanical tests necessary to assess the integrity of the composite material and the structure.

2. NONDESTRUCTIVE EVALUATION (NDE) TECHNIQUES

In the broadest sense, nondestructive evaluation (NDE) includes any examination that assesses material integrity without damaging or destroying a component, and it is used for a variety of tasks such as in-process quality control, in situ test monitoring, and fleet service inspection. For failure analysis, nondestructive evaluations are useful for identifying the conditions of invisible fracture for documentation and for planning subsequent destructive evaluations.

Following preliminary visual and macroscopic analysis of the failed component, invisible damage can be identified and evaluated by various techniques outlined below. Figure 3.1 presents a summary of the various methods commonly used for failure analyses, as related to typical defect or damage conditions. Note that the methods are listed in the order of preference for evaluating and defining each defect condition. The NDE examination (Figure 3.2) is structured such that inspections involve progressively more detail of the damage condition, including

- initial plan view inspections
- detailed plan view inspections
- through-thickness inspections

It should be noted that a majority of the failure analysis investigations do not require detail beyond the initial plan view inspections. Usually, failure of the part denotes that some fairly extensive fracture, often visible, has oc-

TECHNIQUE*	DESCRIPTION	USE	VALUE
Thru-transmission ultrasonic (TTU) • C-Scan	Measures ultrasonic sound attenuation • C-Scan-plan view presentation	Determines size & location of nonvisible damage, defects, fracture in plan view.	• Plan view documentation of failure • Plan view assessment of part quality • Planning analyses
Pulse ultrasonic • B-Scan • C-Scan	Measures ultrasonic sound reflection • B-Scan-thru-transmission view presentation • C-Scan-plan view presentation	Determines size & location of damage, defects, fracture in both a plan and thru-thickness view.	• Plan view documentation of failure • Thru-thickness view documentation of failure • Plan view assessment of part quality • Thru-thickness assessment of part quality • Planning analysis
Reasonance bond testers • Bondascope 2100 • Sondicator • Acoustic flaw detector • MIA 3000	Measures mechanical resonance changes caused by defects, meter, or CRT display	Determines size & location of nonvisible damage.	• Determining of size & location of part damage • Method can be used when only one size is accessible
X-ray radiography (tomography)	Measure X-ray attenuation plan view presentation	• Determines size & location of translaminar fractures & radio-opaque defects - plan view presentation • Delamination size & location determined with radio-opaque penetrant • Thru-thickness position determined by stereo-radiography or X-ray tomography	• Plan-view documentation of failue • Plan-view assessment of part quality & defects • Planning analyses
Neutron radiography	Measures neutron attenuation plan-view presentation	• Determines size & location of translaminar fractures & neutron opaque defects-plan view presnetation • Delamination size & locations determined with neutron-opaque penetrant • Used often where metal structure overlays composite material-neutrons are not as attenuated by metal as X-ray, and are relatively sensitive for polymers with hydrogen.	• Plan-view documentation of failue • Plan-view assessment of part quality & defects • Planning analyses
Eddy current	Measure conditions which interrupt the flow of eddy current induced in the part	Determines differences between paint scratches and surface cracks in Gr/E fabric structures	• Plan-view documentation of fracture with single-size access • Planning analyses

*listed in order of preference, based on applicability, reliability, cost, and sample requirements

FIGURE 3.1. Failure analysis techniques—NDE.

116

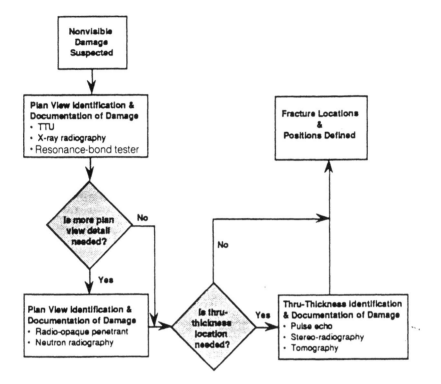

FIGURE 3.2. Nondestructive evaluation.

curred. The main thrust of NDE inspection is to define the damage region around the primary fracture. More detailed analyses, if performed, are usually confined to other regions on the component, away from the principal fracture region, so as to identify other sites of damage or contributory defects. NDE aspects for composites can be found in a number of references [3–16].

The NDE techniques for defect (damage) assessment in relation to structural configuration are summarized in Figure 3.3.

2.1 Initial Plan View Inspections

For preliminary inspection, the major emphasis is to determine the basic outline of the damage regions such that part breakdown and sectioning can be performed without destruction of evidence or to minimize the sectioning if a repair scheme is considered. Plan view analyses such as ultrasonics [through-transmission ultrasonic (TTU) or pulse echo] and radiography are

STRUCTURES	DEFECTS				
	DELAMINATION/ DISBOND	IMPACT DAMAGE	FASTENER HOLE DAMAGE	TRANSLAMINAR FRACTURES, SURFACE	TRANSLAMINAR FRACTURES. SUBSURFACE AND SUBSTRUCTURE
LAMINATED SKIN	C.D.E	A,C,D,E, G	A,C,G,H	A,B	C,G,F, D,E
SKIN/HONEY COMB PANEL	C.E.D.	A,C,E,D	A,C,G,H	A,B	C,G,F, D,E
SKIN-TO-STIFFENER JOINT	C.D.E	A,C,D,E, G	A,C,G,H	A,B	C,G,F, D,E
SKIN-TO-METAL JOINT	C.D.E	A,C,D,E	A,C,H	A,B	C,I,D,E
SKIN-TO-SKIN TRAILING EDGE JOINT	C.E.D.	A,C,E,D,G	A,C,G,H	A,B	C,G,F, D,E
LAMINATED RIBS, SPEARS, AND SKIN STIFENERS	D.E.	D,E,G	A,D,E,G,H	A,B	C,G,F, D,E
LUGS AND THICK SECTIONS	C.D.E	A,C,D,E	A,C,G,H	A,B	C,G,F, D,E

Key:

Inspection Method

A	Visual	F	Fow KV X-ray
B	Penetrant	G	BIB-enhanced X-ray
C	Ultrasonic TTU	H	Eddy current
D	Ultrasonic pulse echo	I	Neutron radiography
E	Bond tester		

FIGURE 3.3. Method selection.

by far the most versatile and encompassing techniques for overall determination of the basic outline of the damage region. Commonly, this coarse
damage assessment is necessary for field inspection prior to part breakdown, sectioning, and subsequent detailed NDE techniques performed in
the laboratory. In the field, pulse echo is the preferred method, which is

particularly desirable in conditions where only one side of the structure is accessible. While this initial inspection can be performed in the laboratory on a fairly flat panel, the TTU C-scan method is by far the most preferable method due to its ability to provide a full-scale, plan view and hard copy record of the defect conditions that are aligned normal to the interrogating beam (delaminations). Defects aligned parallel to the beam direction (translaminar cracking) do not often create appreciable or detectable attenuation and, thus, should be examined by X-ray methods.

2.2 Detailed Plan View Identification

When more plan view details are required, enhanced X-ray and neutron radiography should be used. Each technique can identify both translaminar and delamination damage, although the X-ray technique requires a free edge or surface-intersecting damage so that a penetrant can be introduced. However, when a surface defect is present, enhanced radiography is probably the single most sensitive and accurate inspection technique for composite structures. On the other hand, neutron radiography can be used where metal structure overlays composite material, since neutrons are not attenuated significantly by metal as are X-rays. Also, neutrons are relatively sensitive to polymeric materials containing hydrogen. Neutron radiography, however, has not been proven to generate radiographs that exhibit much contrast or resolution. Another technique available is the eddy current method, which allows for small translaminar cracks to be identified without the requirement of a free edge.

2.3 Through-Thickness Identification

Where through-thickness determination of the location of planar defects such as delaminations are required, either ultrasonics (A-scan, B-scan, or time-domain grated C-scan) or X-ray (stereo radiography) can be used. This provides the investigator with information similar to cross-sectional viewing of the planar defect locations. In Figure 3.4, ultrasonic data presentations for A-, B-, and C-scans are illustrated.

Several techniques such as X-ray tomography, neutron radiography, and resonance/impedance bond testing are available, although their application to failure analyses is extremely limited due either to immaturity, expense, or limitation of field investigation. Holography, acoustic emission, thermography, and speckle photography are not used for post-failure analyses investigations since they require some mechanical loading of the part to define damage states. For the aforementioned reasons, these paragraphs describe in detail the proven and directly applicable NDE techniques such

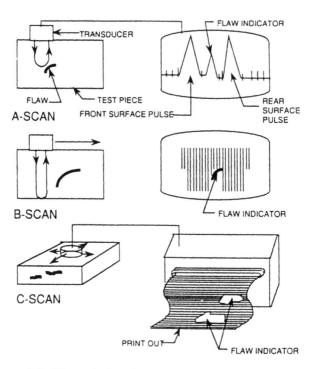

FIGURE 3.4. A-, B-, C-Scan ultrasonic data presentations.

as ultrasonics, X-ray, eddy current, and edge replication. The basic operational modes and uses of each technique is presented, along with some fundamental theory, with regard to evaluating fractured composite structures.

2.4 Evaluation Plan

Nondestructive evaluations can be extremely useful in failure analysis by revealing visible damage, as well as areas of damage not readily discernible through the most intensive visual examination. Nondestructive evaluation is of particular importance with composite materials, primarily because of their susceptibility to invisible internal delaminations within or between the laminate plies. Other identifiable defect or damage conditions found by NDE include

- translaminar surface and subsurface fractures
- core cell damage and fluid ingestion
- porosity
- disbonds

- impact damage
- fastener hole drilling damage
- lightning damage
- heat or fire damage

In addition to its primary benefits, NDE has several other notable advantages. Providing adequate documentation prior to destructive sectioning is of particular importance in failure analysis. With many modern and well-developed nondestructive analysis techniques available for composites (ultrasonics and X-ray being the most common), a permanent record is made of both visible and invisible areas of fracture. This record, somewhat like optical photography, provides invaluable documentation for later perusal. In addition to documentation of the extent of invisible damage, the investigator often gets an added benefit of identifying the type of construction such as locations of core splicing, potting regions, and ply drop-offs or pickups. Finally, since NDE is usually performed by a support specialist, the failure analyst is freed to establish a coherent plan for more detailed evaluation prior to destructive operations (such as specimen removal by sectioning).

To provide the investigator with a better definition for the sequence of steps involved in the nondestructive evaluation of a part, the approach presented in Figure 3.2 was developed by the Air Force. Several goals were considered in creating this approach; first, the chart should provide the most basic information, and second the techniques should progress from the most easily interpreted to the more complex. The chart begins by documenting the fracture in general terms using relatively simple techniques. Next, two operations provide additional plan-view and through-the-thickness information using more complex techniques.

2.5 Ultrasonic Methods

Ultrasonic inspection techniques are useful in characterizing material flaws such as delaminations, cracks, voids, matrix rich pockets, and changes in thickness. For homogeneous materials such as metals, the techniques and the interpretation of the data is well developed and relatively simple. The anisotropic nature of composites presents an added dimension of complexity to ultrasonic inspection.

The variables affecting an ultrasonic high frequency sound source that is directionally focused on the material to be inspected include the acoustic properties of the material and the transporting medium that the beam passes through. Any variation of the acoustic properties of these materials can produce changes in the attenuation (transmission loss), velocity, reflection amplitude, refraction angle phase, and diffraction of the beam. These changes form the basis of various ultrasonic techniques.

The four primary factors that affect the ultrasound transmission in composite structures include: (1) the inherent physical properties of the material, (2) the microstructural features, (3) the condition of the surface, and (4) the thickness of the material. The first and foremost parameters that affect transmission are the physical properties of the material such as stiffness and density, which determine the directions and energy breakdown of the ultrasonic beam within the material. Second, the microstructural features such as resin content, porosity, matrix cracking, delaminations, and ply orientation all affect the ultrasonic sound propagation characteristics. It is the measurement and interpretation of the ultrasonic information that constitutes the major task involved. Figure 3.4 illustrates ultrasonic data presentations.

Typical ultrasonic inspection of homogeneous isotropic materials is a process that requires a trained operator using precision and care to avoid errors in the assessment of the damage conditions. The complexities that arise from analyses of extremely anisotropic materials such as composites requires knowledge of the above-mentioned factors that can have pronounced effects on the information obtained. Although these factors tend to complicate the analyses, they can also contribute in a constructive manner to provide valuable information regarding the microstructure and basic construction of the component.

There are two primary methods that are recognized as the most flexible and efficient methods for obtaining ultrasonic sound propagation data from composite structures, these are the through-transmission ultrasonic, and pulse-echo techniques.

2.5.1 THROUGH-TRANSMISSION ULTRASONIC (TTU)

In the TTU method, an ultrasonic transducer is placed on one side of the material and emits an acoustic pulse. The pulse travels through the material and is received by a second transducer located on the other side of the material. These transducers are placed in axial alignment so that their common axis is perpendicular to the surface of the specimen. With this placement, the amount of energy transmitted through the material is maximized and can be easily monitored as a function of position when the material is scanned by the transducers. For a C-scan, the entire surface is inspected by moving the transducer in a series of closely spaced traverses with a mechanical system. Most mechanical systems only allow planar scanning motions of flat or circular symmetric shapes. Water jet techniques have been developed, which allow inspection of parts too large to be immersed in a tank. Current technology exists for robot-controlled manipulators that track complex surface geometries, but such systems are not yet commonplace. Since the speed of the test is limited primarily by the speed of the scanning, several arrayed transducers are often used for large-scale inspection tasks to reduce scan times; however, for most failure analyses, a single transducer can inspect a large part in a few hours.

The transmitted sound can then be evaluated and broken down into several sublevels, or gray scales, with each level equivalent to a certain amount of attenuation. This energy loss can be related to either a voltage or decibel (dB) metric. Each sublevel can be assigned numbers or colors, and graphically presented as a plan view of the part. Regions of attenuation greater than a standard, say in the 6 to 18 dB range, indicate the presence of significant damage conditions that reflect the energy of the sound beam. Through the use of real-time computer monitoring of the attenuation, a map of the sound transmission relative to the part geometry can be produced by a plate in which an image is formed by burning the surface layers of an ink-imgregnated conductive paper. The numbers denote a range of dB sound loss, with the larger and darker numbers indicating more attenuation.

In TTU, the sound attenuation results from three sources: viscoelastic effects, geometric dispersion due to material anisotropy, and dispersion due to geometric internal damage. By proper selection of the sound frequency, the attenuation due to delamination and cracks can be maximized and the attenuation due to material viscoelasticity and heterogeneity can be minimized. The use of 1 MHz has been recognized to provide the best transmittance since it has a fairly long wavelength and thus is less susceptible to scattering from smaller structural details, particularly for sandwich parts with a honeycomb core. When increased sensitivity to smaller details is desired, the use of smaller wavelengths is used; however, frequencies in this range (5 to 15 MHz) do not transmit through a honeycomb structure and require a more critical alignment of the two transducers.

The transporting medium involved in TT is usually water, which provides a uniform coupler to transmit the sound waves between the transducers and the specimen. This requirement for a water coupler basically limits the TTU inspections to the laboratory; however, a few portable units are available. The specimen is either immersed or water jets at the transducers supply a stream as the coupling agent. With composite failures, surface damage in the form of edge delaminations or translaminar cracking is often present. This surface damage, if extensive, allows water to penetrate into the cracks. Since the attenuation is commonly due to air at the crack impeding the transmittance of sound, the water penetration can displace the air and eliminate, or significantly reduce, the attenuation at these defects. Hence, special precautions are required to prevent the intrusion of water into these areas, particularly for those specimens where a contaminant is suspected and water would be very undesirable. Normally, the open surface cracks are edge sealed with adhesive tape to inhibit entry of water.

Inspection of the TTU C-scan plan view records can provide a full-scale assessment of the major defect conditions. Usually, this inspection method is adequate to define the general outline of the delamination, particularly surrounding the major damage region that is visible. Although the C-scan method is best used to define delaminations, much smaller defects such as porosity can also be identified in extreme cases. This technique is limited by the fact that both sides of the material must be accessible, the depth of defects within the laminate cannot be determined, extreme variations in thickness cannot be evaluated at the same time, and defects aligned parallel to the incident beam are not easily identifiable (such as translaminar cracking). Where situations exist that require a more detailed inspection of the damage such as the determination of defect depth, with access to only one side of the part, the pulse-echo method should be employed.

2.5.2 PULSE-ECHO ULTRASONICS

In the pulse-echo method, a single transducer transmits and receives the acoustic pulse. The transducer emits a gated pulse through the material, which is reflected at the far side of the part and then detected by the transducer again. Since the full dynamic range of the receiver is available to amplify any backscattered acoustic energy, this technique can be made quite sensitive to subtle defect conditions. The reflections from the front and back surface provide known time-related endpoints so that the depth of the defect can be determined by its time function. A potential disadvantage of this method is that flaws one ply away from the front or back surface can be hidden by the reflections from the surfaces. This problem can be alleviated by properly adjusting the instrument to distinguish between these reflections, in combination with using a delay line transducer. Additionally, it is necessary to record the returned echo trace and section it at various periods of time in order to have an accurate representation of the location of the flaws. Breaking down the echo trace allows for the differentiation and separation between closely arranged flaws and prevents the investigator from mistaking several small flaws as a single large one.

For use in the C-scan format, the inspection is usually performed in either a water bath or by using columns of water sprayed upon the surface of the specimen. The water serves as the coupler and delay line for the ultrasonic signals. Information is generally recorded such that the signal levels are graphed or visually displayed on a scope as the transducer is transversed over the specimen. The C-scan pulse-echo is a plan view, that is a two-dimensional image of the internal structure of a material. With gating of the amplitude-based digital signal, imaging of defects can be identified as related to the position within the thickness of the material. The use of a combination of two gating zones can allow the differentiation of delaminations near the front surface (light) and the back surface (dark).

Short pulse (shock wave) methods can also produce a considerable amount of information, although it is limited to linear plotting of the data instead of the two-dimensional C-scan. In A-scan, the reflected pulses can be displayed in real time on a cathode ray tube or permanently recorded. By comparing the reflected pulse information from a region of damage with an area containing no damage (often a calibration sample), the depth of the defect condition can be determined with reasonable accuracy. Similar to C-scan, this method produces an image delineating the reflections between the front and back surfaces. By taking several parallel passes of the transducer over the part, a better feeling of the three-dimensional geometry of the defect can be obtained.

While the A-scan provides data regarding all reflections through the thickness, the B-scan indicates only the first echo after entering the surface. It is therefore incapable of displaying second and higher multiple reflections, as the other two methods can. The B-scan is somewhat similar to a transverse cross section, in that it provides a record of the depth location on a line across the specimen surface. By taking several of these scans, they can also be combined to provide a more complete view of the through-the-thickness damage in the material. For field or initial inspections, the A-scan method using a cathode ray tube for imaging can be used. For instance, early NDE investigations on large failed structures with both visible and non-visible damage used a hand-held pulse-echo scanner to quickly define the perimeter of the damage. For such applications a couplant consisting of a light grease, oil, or a commercially available gel can be used.

2.5.3 SINGLE-SIDED ULTRASONIC (BACKSCATTER)

Recent developments have provided a method of identifying transverse cracking, with resolution capable of identifying cracking within a single ply. This technique uses single-sided ultrasonic angle beams (backscatter) and involves using an off-axis transducer that imparts the signal at an angle to the surface. Also, this technique utilizes separate transmitting and receiving transducers housed in a single surface probe. A complex wavefront is created in the structure by the transmitter and is continuously monitored by the receiving transducer. With this system, delaminations create both a phase and amplitude change, allowing their detection. Because the ultrasonic signal is angled with respect to the surface, this technique has en hanced sensitivity to cracks oriented in the through-the-thickness direction. A drawback is that each ply must be scanned at a specific angle of incidence to maximize the signal amplitude and properly determine the location of the microcracks. A reference standard is also required that duplicates the materials and lay-up of the laminate being inspected.

2.6 X-ray Radiography

X-ray techniques can be used to detect large subsurface fractures when there is a displacement of members. Extensive cracks in which surface displacement has not occurred may not be detectable. Defects such as a crushed core and fractures associated with impact damage lend themselves to radiographic evaluation. In addition to detecting cracks, the investigator can learn valuable information regarding the internal structure of the component, particularly if there are major differences in construction such as stiffeners and honeycomb cores. Water in the core, if extensive, can also be detected. For failure analysis, the use of X-ray radiation is often considered as a secondary experimental technique to, ultrasonics. Although it possesses advantages for situations in which the damage is located parallel to the incident interrogating beam such as translaminar cracking.

2.6.1 CLASSICAL RADIOGRAPHY

An X-ray (radiographic) inspection is performed by transmitting a beam of penetrating radiation through an object onto a photosensitive film. This beam is partially absorbed by the composite as it passes through. Discontinuities such as translaminar cracks cause a reduction in thickness parallel to the incident beam path and, consequently, results in less absorption and less reduction in the intensity of the X-ray beam. These varying beam intensities, which strike the film plane, form a latent image. The film is processed to form a visible image called a radiograph. The radiograph is then evaluated for information regarding the extent and nature of the defect conditions, if they exist.

Although conventional radiography readily detects through-thickness fractures, it does not always present conclusive results. The information obtained from radiographic inspection is affected by the size and orientation of the defect relative to the incident beam. Defects presented normal to the beam result in insufficient changes in density so that interlaminar defects such as delaminations and porosity are not detected by conventional radiography. Due to these limitations, the technique has been significantly improved by the use of some form of radio-opaque penetrant, which probably produces the single most sensitive inspection technique for detecting cracks in composites that are surface related. The following paragraphs focus on this valuable technique.

2.6.2 PENETRANT-ENHANCED RADIOGRAPHY

The procedures for making radiographs of defects and damage in composites differs from the conventional ones in that an X-ray opaque penetrant is used to enhance the damage. The penetrant provides significant improvement in contrast to the damage and the intact composite, as shown in Figure 3.5. Various penetrants have been used, with tetrabromoethane (TBE) being the first solution evaluated. This penetrant was found to be highly toxic and carcinogenic so its use was discontinued. Diiodobutane (DIB) was used for a short time but was also found to be a toxic organic halide, expensive, and had a short shelf life. The enhancement chemical that has proven nontoxic and is relatively inexpensive is zinc iodide (Z_nI_2), used with an alcohol carrier solution. The procedures for use of the carrier solutions are as follows:

Isopropyl Solution	MEK Solution
Zinc Iodine—60 grams	Zinc Iodide—60 grams
Water—10 ml	MEK—250 ml
Isopropyl Alcohol—10 ml	
Kodak "Photoflow"—1 ml	

The isopropyl solution requires approximately thirty minutes after application for the penetrant to encompass the damage zone. The time indicated can be reduced to about ten minutes using a MEK solution. After saturation, excess penetrant should be removed from the open surfaces with an absorbent material. The use of these penetrant materials should be used with caution, since they can potentially damage or obscure the fine microstructural fracture details and prevent detection of contaminants. For chemically stable matrix systems, such as epoxies, the use of these penetrants has not created any undesirable effects such as damage to the fracture surface details, particularly when removed with a clean solution of the primary solvent carrier. For other material systems such as thermoplastics, the effects have not been evaluated and, thus, should be spot tested prior to application to the fracture surfaces.

Various X-ray films can be used, although a high resolution, single-coat film such as Kodak Type R industrial, low speed, fine grain film gives the best contrast and resolution. Double-coated film should be avoided since it is exposed on both emulsions, leading to a double image and loss of resolution.

TECHNIQUE	DESCRIPTION	USE	VALUE
Optical macroscopy	Optical examination at magnification generally at or below 10X	Plan-view examination and identification of fracture surface features, damage and defects	• Documentation of fracture • Identification of fracture types (translaminar vs. interlaminar) • Determination of translaminar fracture modes
Optical microscopy	Optical examination at magnifications above 10X	Plan view examination and identification of fracture surface features, damage, and defects	• Identification of fracture types (translaminar vs. interlaminar) • Determination of interlaminary fracture direction, mode, and environment • Determination of origin • Identification and characterization of defect and damage conditions
Optical X-section	Metallographic optical examination at magnifications above 10X	Cross-sectional examination of laminate, defect, and damage conditions	• Identification of fracture locations • Determination of laminate orientation and drawing compliance • Identification and characterization of defect conditions
Scanning electron microscopy (SEM)	Microscopy performed by mapping; secondary electrons from the sample generated by a primary electron beam raster	High-magnification examination of fracture surfaces and defects with excellent depth of field	• Documentation of fracture surface • Identification of interlaminar fracture mode, direction, and environment • Identification of translaminar fracture mode, direction, and environment • Determination of origin • Identification and characterization of defect conditions

FIGURE 3.5. Failure analysis techniques—fractography.

TECHNIQUE	DESCRIPTION	USE	VALUE
Transmission electron microscopy (TEM)	Microscopy performed by examining the focused pattern of electrons attenuated by a thin fracture surface replica	High-magnification examination of replicated fracture surfaces with better depth of field than in optical microscopy	• Limited to delamination fractures • Documentation of fracture surface • Identification of interlaminar fracture mode, direction, and environment • Determination of origin • Identification and characterization of select conditions
Back-scatter electron microscopy	Microscopy performed by imaging back-reflected primary beam electrons generated by a rastered electron beam	Intermediate magnification of fracture surface and defects. Back-reflected electrons are sensitive to atomic number and can be used to distinguish surface details as a function of atomic number.	• Documentation as a function of atomic number • Identification and characterization of defect conditions

FIGURE 3.5 (continued). Failure analysis techniques—fractography.

X-ray units emitting soft X-rays are recommended, with a small spot size in the range of 1.5 mm by 1.5 mm or less, with an inherent filtration of 1.0 mm beryllium equivalent or less. The generator tube should be capable of producing a minimum of 20 kilovolts at 2 milliamps. The low operating voltages produce soft X-rays that provide resolution of structural details within the laminate, such as porosity and fiber spacing irregularities. For applications that require positioning the X-ray tubes at tight locations, the use of an end anode side emission is recommended. The optimum exposure times are those that produce high resolution negatives from which prints can be made. These exposure times are shorter than that used for direct viewing. It should be noted that it is difficult to obtain prints of radiographs that will reproduce adequately by normal printing methods, and therefore reporting and presentation of the results are more difficult and somewhat lacking.

Inspection of the films often requires an expert to differentiate artifacts from actual damage. The interpretation requires an understanding of how the penetrant affects the X-ray beam and how it enters the damaged specimen. Regions containing penetrant appear darker than regions containing no damage or defects. Regions containing no penetrant have a uniform grayness, especially on structures that have small thickness variations, since composites have relatively low radiographic scattering or absorbance due to the elements present. Cracks such as matrix or translaminar crack-ing appear on the film as long, narrow, dark lines. The interpretation of ar-tifacts corresponding to delaminations is usually more difficult. The open-ing displacement of the crack is greatest at the edge and least at the end of the delamination, therefore, one might expect a visible lightening of the im-age from the free edge toward the crack tip. While this change in gray level holds true at the extreme ends of the crack, the situation at the central re-gion is such that the capillary forces are not strong enough to hold the pene-trant. This results in a central boundary region that does not stop the X-rays and often appears light and undamaged.

Another modification to the penetrant enhanced X-ray image involves making stereo radiographs, such that a three-dimensional view of the inter-nal damage can be examined. The standard stereo radiography procedure consists of making two X-ray films of an object from slightly different ori-entations. The best method for creating the two views is to rotate the part through a small angle (usually about seven to fifteen degrees). The part is then allowed to remain at the center of the path of the X-rays and is also centered in the radiograph. The depth of the damage can then be identified, as well as differentiating between overlaying damage that might be masked with a single plan view. With the aid of a stereo viewer, the defects nearest the X-ray source have the largest relative displacement, and the furthest defects are the least displaced.

2.7 Eddy Current

This technique is commonly used in metals and has provided satisfactory damage interrogation for fabric laminates, particularly for locations around small, localized geometric variances such as fastener holes and edge radii. Eddy current testing involves the use of small, hand-held surface probes that produce an alternating magnetic field. The magnetic field is generated by an alternating current test instrument coil. This alternating expanding and collapsing current induces a magnetic eddy current in the specimen. The interaction of this magnetic field with the test instrument varies as in-ternal flaws and fractures are encountered. The use of this instrument is basically limited to solid laminates that are conductive and have appreciable magnetic permeability. An example of the use of this method is for carbon fiber composites. This method relies on the conductivity of the carbon fiber, which at best, are limited.

2.8 Edge Replication

Edge replication has proven itself to be an accurate technique for docu-menting the state of damage in thin laminate sections. It is a direct application of the replication technique used for TEM specimen preparation. An acetate film that has been softened with acetate solution is firmly applied to the edge, then allowed to dry. The replica can then be shadowed to enhance the surface features. The result is a mirror image of the edge that can be examined at higher magnification to assess invisible damage. Cracks such as translaminar cracks for 90 degree plies and edge delaminations in 0 degree plies can be readily identified and highlighted by this technique.

3. DAMAGE ASSESSMENT: OPTICAL/SEM EVALUATION

3.1 Optical Microscopy

Optical microscopy has proven itself a most critical tool for failure analyses, for the examination of both fracture surfaces and cross sections. For cross sections, the optical microscope remains the single most important technique available. The analyses of cross sections provide insight into the microstructural features related to construction, crack propagation, and defect conditions of the composite. For fracture surfaces, particularly delamination surfaces, the optical microscope is possibly the best technique and, at the least, should be used in conjunction with the SEM and TEM, rather than as a stand alone. The fractographic depiction of delaminations by optical microscopy at fracture, the origin locations, and anomalous conditions are related to the origin. Undoubtedly, this information is considered paramount to the determination of the cause and sequence of failure and thus should be required for all investigations of delamination surfaces.

3.1.1 SPECIMEN PREPARATION

A reflected light illumination mode requires a relatively flat surface due to depth of field limitations, and therefore specimen preparation is important to provide the best situation for examination and documentation. Specimen preparation for fracture surface examinations involve cutting the desired fracture region from the remaining structure followed by cleaning.

3.1.2 FRACTURE SURFACE INSPECTIONS

Optical fractography is by far the most efficient and cost-effective method for examination of delamination surfaces. Since the specimen setup and examination times are very short, a large area of fracture surface is covered using this technique. As a result, a reliable and accurate determination of the typical and representative fracture surface features is obtained in a relatively quick fashion. Translaminar fractures, on the other hand, are rough and have features (fiber ends) that do not readily lend to optical microscopy. Low magnification inspections can be used on delaminations to verify the plane of fracture and the location of crack growth features such as bench marks and transverse cracking. More detailed, higher magnification inspections provide resolution of the fine matrix resin fracture features. Bright field illumination, a stepped-down aperture, and long working distance objective lenses provide the best combination for examination at high magnifications as required for crack mapping and identification of the fracture modes. The features found with the optical microscope, even though they are visible through the eyepiece, are often too small to document with photographic film. In these situations where photomicrographs are desired, the SEM is required.

3.1.3 CROSS SECTION ANALYSES

Metallographically prepared cross sections provide the following types of information:

- determination of the overall laminate quality
- quantitative evaluation of the extent of porosity, relative percent and morphology of phases or microconstituents, and ply count and orientations
- origin examinations
- inspections of interfacial conditions
- crack propagation regions (intraply versus interply)
- extent of degradation due to wear, thermal cycling, and fatigue

Where possible, specimens should be selected from at least two areas to most accurately characterize quality features of the composite, particularly when anomalous conditions are identified. These two areas should be selected such as to be identified (1) as close to the origin as possible and (2) at an area away from damage.

Several illumination methods are available for cross section analyses, with bright-field being the most widely used. Dark-field or oblique illumination provides an excellent image contrast for differentiating surface topographical features such as microcracks and phase interfaces. Polarized light can be used to enhance differences between ply orientations for easier ply count and orientation analyses.

3.2 Scanning Electron Microscopy (SEM)

Of relatively recent origin, the Scanning Electron Microscope (SEM) has found a wide range of applications in failure analysis, materials research and development, and quality control. Fractography is probably the most popular application of the SEM. The three-dimensional appearance of SEM fractographs, with large depth of focus, large magnification range, and simple specimen preparation with direct inspection make the SEM a versatile and indispensable tool in failure studies and fracture mechanism research. This unique instrument offers possibilities for image formation of fractured parts that are usually easy to interpret and reveal clear photomicrographs of rough surfaces, as well as polished cross sections. The development of an assortment of related capabilities, such as stereo viewing, quantitative microchemical analyses, in situ fracture studies, and image formation that is easy to interpret, all contribute to the value of this investigative and research tool. Energy dispersive X-ray (EDX) analysis equipment is routinely attached to the SEM, providing semiquantitative and, in favorable situations, quantitative analysis of composition from a small volume. For composites, EDX analysis is usually only required for contamination analysis as presented in a previous section.

The SEM is capable of magnifications over a range extending from 5 × to 250,000 ×, although the majority of composite fractography analyses do not generally exceed 20,000 ×. The SEM has a normal resolution of approximately 100 angstroms. The depth of field is about 300 times that of the light optical microscope, providing an excellent three-dimensional view of the specimen at focal depths of over 1,000 microns at 100 × and 5 microns at 20,000 ×. Specimens can be tilted up to approximately 70 degrees to the incident beam, for specimen examination while maintaining focus over an extremely rough surface (such as with protruding fibers). The specimen size is usually limited by the size of the chamber door. The maximum size for the latest SEM equipment is approximately 6 in. by 6 in., with limitations in tilt; thicker specimens and maximum tilting can be accomodated. Since the specimen size is limited, very large specimens must be partially destroyed. There is a limiting feature to the SEM, and therefore lower magnification and less destructive inspections such as optical macroscopy and microscopy should be employed prior to SEM analyses.

A summary of failure analysis techniques are presented in Figure 3.5.

4. DAMAGE TOLERANCE TESTS

Damage tolerance evaluation for composites of different classes of materials are not well defined except for advanced polymeric composites used in aerospace applications. Damage tolerance is a safety issue, and, because of the extensive use and experience with carbon fiber polymeric composites and concern for safety, data required to make assessments on damage tolerance characteristics have become mature. For metal matrix composites and ceramic matrix composites, this is not the case. These technologies are evolving, and application of these materials are yet to be introduced in aerospace applications in greater volumes. Because of the lack of an experience base for these material systems, damage tolerance requirements have not as yet been properly defined. Also, appropriate tests have not been defined and established. Failure modes for these composites are substantially different such that the methodology developed for PMCs is not easily transferable. For glass fiber composites (e.g., automotive and marine industries), typical applications have been in areas where safety issues have not been as critical as in the aerospace area. Therefore, there is a need for a damage tolerance methodology for this class of composites.

Even in the aerospace industry, based on applications, requirements for damage tolerance may vary. However, there is a consensus that low-velocity impact appears to be a major concern for PMCs and defines the event for which damage tolerance design should be based. Compression after impact (CAI) tests have become the cornerstone test as an evaluation technique to assess the damage tolerance of PMCs. This type of test and a set of other relevant tests have been documented in the NASA-1092 document. In estimating the damage tolerance capability of PMCs, these tests provide critical information on the structural capability of the material. These tests are described in the following sections.

4.1 Compression after Impact Test

As an example of a compression after impact test, a graphite/epoxy test laminate having a nominal thickness of 0.25 in. (6.35 mm) and an orientation of $[+45/0/-45/90]_{ns}$ is considered. Usually an ultrasonic inspection to determine the quality of the laminate is performed prior to testing to failure.

The impact test specimen dimension is given in Figure 3.6. After impact, the specimen is trimmed to a width of ± 0.03 in. (0.76 mm) for evaluating the compressive failure strength.

The impact test apparatus (Figure 3.7) consists of a base plate, a top plate, and an impactor. The impactor weighs 10.0 lb (4.55 kg) and is usually less than 10 in. (25.4 cm) in length and has a 0.5-in. (12.7-mm) hemispherical steel tip on the end that impacts the target. A tube lined with DuPont Teflon, or equivalent, is used for guiding the impactor with minimum friction to the target.

The compression test apparatus provides a simple support to the compression test specimen along its long edges which are oriented parallel to the compression loading direction. The short edges (loaded edge) are clamped between two adjustable steel plates at the upper and lower sections of the apparatus in order to provide resistance to end brooming.

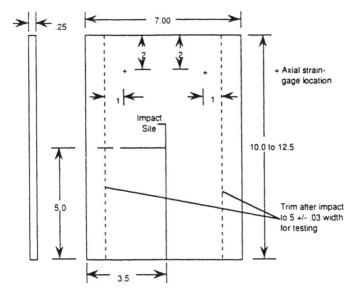

FIGURE 3.6. Compression after impact test specimen (dimensions are in inches).

The graphite/epoxy test specimen is placed in the impact test assembly with the identification side up so that the desired impact location is centered within the 5.0 by 5.0-in. (12.7 cm) central opening in the base plate. The top plate is then placed upon the test specimen and clamped to the base plate by installing nuts on the four tie-down studs and torquing each one to a nominal 20 ft-lb (2.77 kg-m). Alignment pins are provided in the base plate so that the top plate is correctly positioned. The guide tube is located above the test specimen so that the impactor will strike the center of the specimen. The lower end of the guide tube is approximately 10 in. (25.4 cm) above the surface of the specimen. The striker end of the impactor is coated with white chalk dust or white grease to allow easy location of the actual impact point. The impact tup is dropped from a height of 2 ft (50.8 cm) above the test specimen to generate an impact energy of 20 ft-lb (2.77 kg-m). Care should be taken to deflect the impactor away from the specimen after the strike so that a rebound strike does not occur. The impacted specimen is then removed from the test apparatus, and the amount of damage to the specimen on the impacted surface (side with the specimen identification) and the back surface is visually determined. The specimen is then ultra-sonically inspected to determine the extent of internal delamination.

The compression test specimen is then loaded to failure by using a stroke-controlled testing machine. A loading rate of 0.05 in./min (1.27 mm) is recommended. The load-strain behavior of the test specimen is recorded

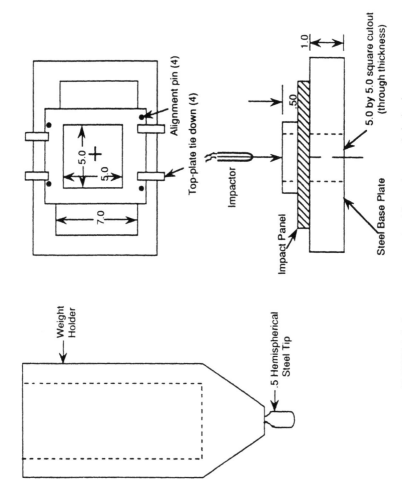

FIGURE 3.7. Impact test apparatus (dimensions are in inches).

throughout the test by using strain gages. The specimen is installed in the compression test apparatus (Figure 3.7) such that (1) the specimen is parallel to the load axis of the machine and is centered in the machine; (2) the side supports on the edges parallel to the loading axis shall be a snug fit, but not tight, so that the specimen can still slide in the vertical direction; and (3) a 0.050-in (1.27-mm) clearance is provided between each side of the specimen and the side supports to prevent any transverse load due to Poisson deformation. Three impact tests are conducted at an impact energy of 20 ft-lb (2.77 kg-mm). The three specimens are then tested to failure in compression.

For each impact test, data is recorded as specified in Table 3.1 where h represents the thickness of one ply of the laminate. Also, ultrasonic measurements using C-scan, and, where possible, associated B-scan records are provided.

4.1.1 EDGE DELAMINATION TENSION TEST

Two panels for selected toughened-resin composite tests are needed. One panel consists of an 11-ply lay-up $[(\pm 30)_2/(90)_3/(\mp 30)_2]$, while the other panel consists of an eight-ply lay-up $[+35/-35/0/90/90/0/-35/+35]_T$. Quality assurance C-scans are performed on both panels. Each test specimen is nominally 10 in. (25.4 cm) long and 1.5 in. (3.8 cm) wide. Other details of the specimen are shown in Figure 3.8.

Specimens are mounted in a properly aligned load frame. Either a stroke-controlled, screw-driven machine, or strain controlled hydraulic machine can be used. (Note: "Stroke-controlled" controls crosshead displacement, whereas "strain-controlled" controls displacement over the gage length of the strain-measuring device mounted on the specimen.) Distance between grips is taken as 7 in. (17.8 cm). Emery cloth or tungsten carbide grit inserts are sufficient to prevent slip at the grips. However, if end tabs are used, they should be squared off, not tapered. Either (1) a pair of LVDTs (linear variable differential transducers) or DCDTs (direct-current differential transducers), one on either side, or (2) an extensometer (clip gage) with an appropriate extended arm can be mounted on the specimen. Specimen gage length is 4 in. (10.16 cm) with gage mounts located at 1.5 in. (3.8 cm) from either grip.

Five specimens of each laminate should be tested. Specimens are loaded at a slow rate [approximately 0.0001 in./sec (0.00025 mm/sec)]. Output of the LVDTs (average of front and back) or the extensometer along the *X*-axis, and load along the *Y*-axis of an *x–y* plotter (real-time analog display) is recorded. Loading is continued until visible detection of edge delamination and for a corresponding abrupt (not continuous) deviation in the load-deflection plot occurs (Figure 3.9). The strain level at the onset of

TABLE 3.1. ST-1 Compression after Impact Test Data.

Company Affiliation: _____

Material: _____ h = _____ mils/ply

Laminate Orientation: [+45/0/−45/90]$_{ns}$
Resin Content: _____ % by weight
Test Condition: 75°F dry

Specimen ID	Thickness, in.	Width, in.	Impact Energy, ft-lb	Maximum Width of Impact Damage, in.	Visual Impact Damage		Impact Area, in.²	Failure Load, kips	Failure Stress, ksi	Failure Strain, μin./in.	Compression Modulus, psi
					Front surface	Back surface					
	x.xxxx	x.xxx	xx.	x.xx	Yes	No	x.xx	xx.xx	x.xx	xxxxx.	x.xx × 10⁶

138

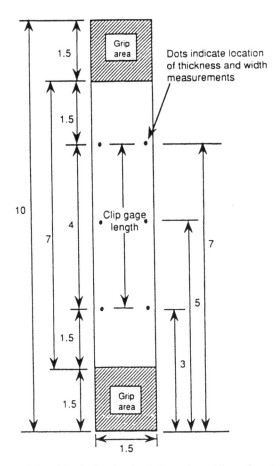

FIGURE 3.8. Edge delamination tension test specimen (dimensions are in inches).

delamination ϵ_c is recorded. The initial laminate modulus E_o is estimated from the linear portion of the load deflection curve. For the laminate shown, the load-deflection plot is linear up to the onset of delamination. However, if gradual nonlinearity precedes the onset of delamination, a secant modulus E_{sec} from the origin to the delamination onset point in the load-deflection plot is measured. This can be of concern for the eleven-ply lay-up $[\pm 30/\pm 30/90/90]_s$ only and not for the eight-ply lay-up $[\pm 35/0/90]_s$. A few of the $[\pm 35/0/90]_s$ specimens are loaded until the specimen fractures into two pieces.

For specimen tests, (1) laminate thickness t; (2) laminate modulus E_o and E_{sec} if necessary; (3) delamination onset strain ϵ_c; and (4) strain at failure

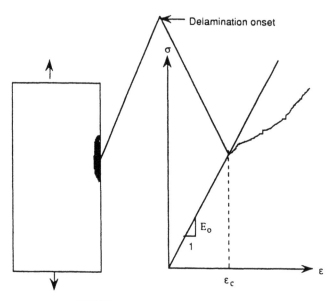

FIGURE 3.9. Critical G_c determination.

for $[\pm 35/0/90]_s$ specimens is recorded. Also, E_{11} and E_{22}, Poisson's ratio ν_{12}, and shear modulus G_{12} for the toughened-resin composite are either known or recorded. These results often come from [0] and $[45]_{2s}$ tension tests. Edge delamination test data as specified in Table 3.2 are recorded.

For the $[\pm 30/\pm 30/90/90]_s$ and $[\pm 35/0/90]_s$ laminate stiffness, E_{lam} are estimated from lamina properties. These results are compared to the average measured laminate modulus E_o. The $[\pm 30]_s$ and $[\pm 35/0]_s$ laminate stiffnesses are estimated from lamina properties; and E^* is then calculated for the eleven-ply lay-up as,

$$E^* = \frac{8E_{[\pm 30]_s} + 3E_{[90]}}{11} \tag{1}$$

while for the eight-ply lay-up,

$$E^* = \frac{6E_{[\pm 35/0]_s} + 3E_{[90]}}{8} \tag{2}$$

The interlaminar fracture toughness G_c is then calculated for each test and then averaged

$$G_c = \frac{\epsilon_c^2 t}{2} (E_{lam} - E^*) \tag{3}$$

where *t* is the average laminate thickness or the thickness measured closest to the delamination if thickness variations are greater than 3 mils (0.08 mm).

Test data are reported in Table 3.2.

4.2 Open-Hole Tension Test

The graphite/epoxy test laminate with a nominal thickness of 0.25 in. (6.35 mm) has an orientation of $[+45/0/-45/90]_{ns}$. Laminate quality is once again checked via ultrasonic inspection.

The tension test specimen width is 2.00 in. (50.8 mm). The minimum length of the specimen is 12.00 in. with a minimum of 8 in. (20.3 cm) between grips. The hole diameter is 0.25 in. (6.35 mm) and is located as shown in Figure 3.10.

The test specimen is loaded to failure at a loading rate of approximately 20,000 lb/min. (9091 kg/min). The failure strength of the specimen is typically in the range of 15,000 to 25,000 lb (6818 to 11,363 kg). Test data are reported as in Table 3.3.

4.3 In-plane Open-Hole Compression Test

Graphite and epoxy test laminates, have a nominal thickness of 0.25 in. (6.35 mm), has an orientation of $[(+45/0/-45/90)]_{ns}$. The compression test specimen (Figure 3.11) has a width of 5.00 ± 0.03 in. (12.7 cm ± 0.08 cm) and length of the specimen is 10.0 in. (25.4 cm) minimum to 12.5 in. (31.75 cm) maximum. The hole diameter is 1.0 in. (2.54 cm) and is located at the center of the specimen. Each specimen has a minimum of four axial strain gages mounted back to back at the locations shown in Figure 3.11.

The compression test apparatus provides a simple support to the compression test specimen along its long edges oriented parallel to the compression loading direction. The short edges (loaded edge) are clamped between the two adjustable steel plates of the upper and lower sections of the apparatus to provide resistance to end brooming.

The compression test specimen is loaded to failure by using a stroke-controlled testing machine. A loading rate of 0.05 in./min (1.27 mm/min) is recommended. The load-strain behavior of the test specimen is recorded throughout the test using all four strain gages. The specimen is installed in the compression test apparatus such that (1) the specimen is parallel to the load axis of the machine and is centered in the machine; (2) the side supports on the edges parallel to the loading axis shall be a snug fit, but not tight, so that the specimen can still slide in the vertical direction; and (3) a

TABLE 3.2. Edge Delamination Tension Test Data.

Company Affiliation: _____

Material: _____   h = _____ mils/ply

Laminate Orientation: [±35/0/90]
Laminate Resin Content: _____ % by weight
Test Condition: 75°F dry

E_{lam} x.xx × 10⁶ psi $E^* = x.xx × 10^6$ psi

$E_{[±35/0]_s} = x.xx × 10^6$ psi

Specimen ID	Thickness, in.	Width, in.	Delamination Onset Strain, μin./in.		Failure Strain, μin./in.	Tensile Modulus, psi	Interlaminar Fracture Toughness, G_c, in.-lb/in.²
			1	2			
	x.xxxx	x.xxx	xxxx.	xxxx.	xxxxx.	x.xx × 10⁶	x.xxx
Average:							

$E_{11} = x.xx × 10^6$ psi $E_{22} = x.xx × 10^6$ psi $G_{12} = x.xx × 10^6$ psi $\nu_{12} = 0.xxx$ where xx etc. are test data

TABLE 3.2. *(continued)*.

Laminate Orientation: $[\pm 30/\pm 30/90/\overline{90}]_s$ E_{lam} x.xx $\times$ 10⁶ psi E^* = x.xx $\times$ 10⁶ psi
Laminate Resin Content: _____% by weight
Test Condition: 75°F dry $E_{[35/0]_s}$ = x.xx $\times$ 10⁶ psi

Specimen ID	Thickness, in.	Width, in.	Delamination Onset Strain, µin./in.		Tensile Modulus, psi	Interlaminar Fracture Toughness, G_c, in.-lb/in².
			1	2		
	x.xxxx	x.xxx	xxxx.	xxxx.	x.xx $\times$ 10⁶	x.xxx
Average:						

1. Strain at first deviation from linear stress-strain curve.
2. Strain at first visible delamination.

143

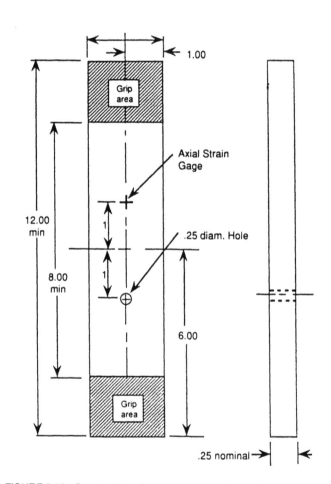

FIGURE 3.10. Open-hole tension test specimen (dimensions are in inches).

TABLE 3.3. *Open-Hole Tension Test Data.*

Company Affiliation: _____

Material: _____ h = _____ *mils/ply*

Laminate Orientation: $[+45/0/-45/90]_{ns}$
Resin Content: _____ % by weight
Test Condition: 75°F dry

Specimen ID	Thickness in.	Width, in.	Hole Diameter, in.	Failure Load, kips	Failure Stress, ksi	Failure Strain, μin./in.	Tensile Modulus, psi
	x.xxxx	x.xxx	x.xxxx	x.xx	x.xx	xxxx.	$x.xx \times 10^6$

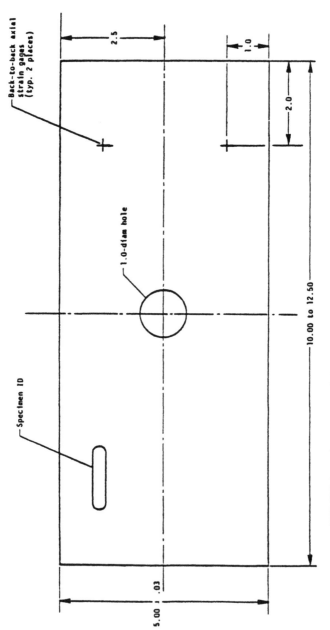

FIGURE 3.11. In-plane open-hole compression test specimen (dimensions are in inches).

TABLE 3.4. Open-Hole Compression Test Data.

Company Affiliation: _____

Material: _____ h = _____ mils/ply

Laminate Orientation: $[+45/0/-45/90]_{ns}$
Resin Content: _____ % by weight
Test Condition: 75°F dry

Specimen ID	Thickness in.	Width, in.	Hole Diameter, in.	Failure Load, kips	Failure Stress, ksi	Failure Strain, μin./in.	Compression Modulus, psi
	x.xxxx	x.xxx	x.xxx	x.xx	x.xx	xxxx.	$x.xx \times 10^6$

147

0.050-in. clearance is provided between each side of the specimen and the side supports to prevent any transverse load due to Poisson deformation.

4.3.1 TEST DATA REPORTING

For each compression test, data, as specified in Table 3.4 where h is the thickness of one ply of the laminate, are recorded.

5. REFERENCES

1. 1987. Boeing Report AFWAL-TR-86-4137.
2. 1983. "Standard Test for Toughened Resin Composites," *NASA*, Washington, DC: National Aeronautics and Space Administration Reference Publication 1092.
3. Henneke, E. G. and J. C. Duke. 1979. "A Review of the State-of-the-Art of Nondestructive Evaluation of Advanced Composite Materials," technical report prepared under Union Carbide Corp. Contract 19X-3673V, Blacksburg, VA: Virginia Polytechnic Institute and State University.
4. Krautkramer, J. and H. Krautkramer. 1977. *Ultrasonic Testing of Materials*, 2d Ed., Berlin, Germany: Springer-Verlag.
5. Papadakis, E. P. 1966. "Ultrasonic Diffraction Loss and Phase Change in Anisotropic Materials," *Journal of the Acoustical Society of America*, 40(4):863–876.
6. Clarke, B. 1990. "Nondestructive Evaluation of Composite Materials," *Inspection and Testing*, (March) pp. 135–139.
7. Lloyd, P. A. 1989. "Ultrasonic System for Imaging Delaminations in Composite Materials," *Ultrasonics*, 27(January):8–18.
8. Buynak, C. F., T. J. Moran and R. W. Martin. 1989. "Delamination and Crack Imaging in Graphite/Epoxy Composites," *Materials Evaluation*, 47(April):438–441.
9. Buynak, C. F. and T. J. Moran. 1987. "Characterization of Impact Damage in Composites," *Review of Progress in Quantitative Nondestructive Evaluation*, 6B:1203–1211.
10. Girshovich, S. 1988. "Nondestructive Testing of Composite Materials," *NDT International*, 21(6):457.
11. Smith, B. T., J. S. Heymen, A. M. Buoncristiani, E. D. Blodgett, J. G. Miller and S. M. Freeman. 1989. "Correlation of the Deply Technique with the Ultrasonic Imaging of Impact Damage in Graphite/Epoxy Composites," *Materials Evaluation*, 47(December):1408–1415.
12. Frock, B. G., R. W. Moran, T. J. Moran and K. D. Shimmin. 1987. "Imaging of Impact Damage in Composite Materials," *Review of Progress in Quantitative Nondestructive Evaluation*, D. O. Thompson and D. E. Chimenti, eds., New York: Plenum Press, 7B:1093–1099.
13. Fatemi, M. and A. C. Kak. 1980. "Ultrasonic B-Scan Imaging: Theory of Image Formation and a Technique for Restoration," *Ultrasonic Imaging*, 2:1–47.
14. McRae, K. I. 1989. "High Resolution Ultrasonic Imaging of Disbonds in Adhesively

Bonded Composite Structures," DREP Technical Memorandum 89-06, Victoria, BC, Canada: Defense Research Establishment Pacific, (March).

15. Bar-Cohen, Y. and R. L. Crane. 1982. "Acoustic Backscattering Imaging of Subcritical Flaws in Composites," *Materials Evaluation*, 40(8):970–975.

16. 1992. "Damage Detection in Composite Materials," *ASTM STP 1128*, J. E. Masters, ed.

Advanced composite materials, 1
Aerostructural, 1
Aircraft fluids, 15
Anisotropic, 41
Axial modulus, 33

Birth defects, 113–114
Buckling limit, 47

Ceramics, 3
Classical radiography, 126
Coefficient of restitution, 106
Complex functions, 101
Composite
    advanced structural, 1
    braided, 2
    brittle composites, 48
    continuous fiber laminated, 2
    flake, 2
    particulate, 2
    random-fiber, 2
    toughened, 18
    tough-resin, 48
    woven laminar, 2
Compressive failure mode, 32
Coupon screening, 46
Crack coupling, 58
Critical strain energy release rate, 95
Critical stress intensity factor, 95
Cyclic loading, 44

Damage
    area, 28
    assessment, 131
    complexity, 14
    containment, 14
    global, 57
    growth, 39
    local, 57
    mechanics, 58
    shape, 39
    size, 7, 39
    state, 39
    splitting, 92
    subcritical, 9
    thresholds, 94
    time, 39
    visibility, 64
Damage tolerance flaw, 47
Damage-tolerant, 44
Damage tolerant design, 47
Defects
    manufacturing, 46
    service, 46
    skin, 46
Deformation mechanics, 58
Degree of damage, 46
Delamination, 14
Dents, 39
Dirac delta function, 79
Dorey's analysis, 21

Ductile, 47
Durability, 37
Durability flaws, 47

Eddy current, 130
Edge delamination, 42
Edge delamination tension test, 137
Edge replication, 130
Effective flaw length, 96
Empirical, 57
Energy
    bending, 68
    contact, 66
    membrane, 69
    shear, 68
    strain, 88
Energy-balance model, 65
Environmental effects, 46
Evaluation plan, 120

Fabrication flaws, 48
Failure-strain, 48
Fail-safe, 5, 8
Failure mode, 10
Fiber
    breaks, 37
    fracture, 12, 58
    hollow, 37
    microbuckling, 19
    miscollimation, 39
    reinforcements, 3
    waviness, 39
    wrinkles, 39
Fiber/matrix debonding, 12, 19
First generation materials, 18
Flexural strain, 27
Flexural test, 22
Foreign
    contamination, 39
    inclusions, 39
    particles, 39
Foreign object impact damage, 15
Fracture analysis, 45
Fracture surface inspections, 131
Fracture toughnesss, 48

Hertzian contact effects, 53
Heterogeneous, 41
Hot/wet deterioration, 29

Impact
    compression after, 12, 134
    damage, 4
    drop weight, 17
    energy, 18
    low velocity transverse, 10
    through penetration, 18
Incident energy, 21
Indentation, 58
In-field/service problems, 114
In-plane open-hole compression test,
        141
Interfacial debonding, 58
Interlaminar defects, 37
Interlaminar shear strength, 20
Interlaminar stresses at joints, 39
Interleaf material, 30
Iosipescu fixture, 22

Kinematic relations, 75
Knee, 21

Lagrangian equations, 78
Laminate
    strains, 103
    stresses, 103
Linear elastic failure, 21
Linear elastic modeling, 21
Loading rate, 41
Load-by-load growth, 45
Local
    failure, 47
    rigidity, 18
    stiffness, 47

Material damping, 65
Matrices
    new composite, 23
    toughened, 23
Matrix
    cracking, 57
    materials, 3
    microcracking, 19
    yielding, 19
Membrane stiffness, 66
Microbuckling failure mode, 32
Mixed mode, 46
Mode I, 46

Modulus
    bending, 21
    flexural, 27
    inplane, 21
    transverse tensile modulus, 28
Multiphase systems, 22

Neat resin ultimate strain, 25
Newmark implicit integration
        scheme, 79
Newtonian mechanics, 72
Nondestructive evaluation, 115
Numerical, 57

Out of plane loads, 37
Open-hole tension test, 141
Optical microscopy, 131
Orthotropic plate compliances, 108

Partial structure failure, 9
Penetrant-enhanced radiography, 127
Penetration, 21
Plan view analysis, 117
Planar voids, 46
Plastic deformation, 21–22
Ply gaps, 37
Porosity, 37
Preliminary inspection, 117
Prepreg, 29
Pseudo-isotropic solids, 63
Pulse-echo ultrasonic, 124

Repeated loads, 44
Residual strength degradation, 58
Resin chemistry, 25
Resin-rich, 37
Resin-starved, 37
Resin strain, 24
Rayleigh-Ritz method, 77

Safe crack growth, 44
Scratches, 14
Semi-empirical, 57
Service loads, 12
Shear crippling, 48
Shear delamination, 42
Shear failure, 19
Shear stiffness, 66
Simpson's rule, 79

Single-sided ultrasonic, 125
Single-phase systems, 22
Slow crack growth, 6, 8, 11
Sources of damage, 38
Specific energy level, 103
Specimen preparation, 131
Spectrum effects, 46
Splitting, 29
Spring-mass model, 69
Stacking sequence, 63
Strain-energy release rate, 45
Strain-displacement relations, 88
Strength
    compression, 10
    flexural, 20, 27
    hot/wet compressive, 15
    residual, 103
    static residual, 9
    tensile residual, 13
    transverse tensile, 28
Stress waves, 41
Striker, 21
Structural dynamic response, 53
Structural integrity, 10, 115
Structures
    primary, 14
    secondary, 14
Surface friction, 65

Target, 40
    curvature, 64
    preload, 64
    reactive force, 68
    thickness, 63–64
Tensile stress, 42
Thermal expansion coefficient, 12
Thermomechanical, 12
Thermoplastics, 3, 15, 100
Thermosets, 3, 15, 100
Thermostructural, 12
Through thickness identification, 119
Through-thickness ultrasonic, 119
Toughened systems, 22
Translaminar
    cracking, 119
    damage, 119
Transverse tensile
    failure, 19
    strain, 28

Triaxial stress, 63–64
Two-spring mass model, 81

Ultrasonic methods, 120

Variational methods, 66

Weight saving, 17
Wet glass transition temperature, 24
Work-to-break, 27

X-ray radiography, 126
Zero-defect, 47

Milton Keynes UK
Ingram Content Group UK Ltd.
UKHW031133141024
449569UK00006B/215